R...   ...ES

Plants of
VANCOUVER
and the LOWER MAINLAND

COLLIN VARNER

RAINCOAST BOOKS
Vancouver

Text and illustrations copyright © 2002 Collin Varner
Photographs copyright © Collin Varner 2002, except for pages 7, 24, 34, 35 right, 36, 39, 40, 56, 75, 85 left, 88, 89 left, 91, 94, 125 centre, 126, 128 copyright © 2002 Greg Thrift

Edited by Scott Steedman and Simone Doust
Designed by Ingrid Paulson

All rights reserved. No part of this publication may be reproduced or transmitted in any form or by any means, electronic or mechanical, including photocopying, recording or by any information storage and retrieval system now known or to be invented, without permission in writing from the publisher.

Raincoast Books
9050 Shaughnessy Street
Vancouver, British Columbia
Canada V6P 6E5
www.raincoast.com

Raincoast Books is a member of CANCOPY (Canadian Copyright Licensing Agency). No part of this publication may be reproduced, stored in a retrieval system or transmitted in any form or by any means without prior written permission from the publisher, or, in case of photocopying or other reprographic copying, a license from CANCOPY, One Yonge Street, Toronto, Ontario, M5E 1E5.

Raincoast Books acknowledges the ongoing financial support of the Government of Canada through The Canada Council for the Arts and the Book Publishing Industry Development Program (BPIDP); and the Government of British Columbia through the BC Arts Council.

NATIONAL LIBRARY OF CANADA CATALOGUING IN PUBLICATION DATA

Varner, Collin.
 Plants of Vancouver and the Lower Mainland

 Includes bibliographical references and index.
 ISBN 1-55192-479-X

 1. Botany–British Columbia–Lower Mainland. I. Title.
QK203.B7V375 2002   581.9711'33   C2001-911683-7

Printed and bound in Hong Kong, China by Book Art Inc.,Toronto

1 2 3 4 5 6 7 8 9 10

TABLE OF CONTENTS

A Great Natural Garden	7
Flowers	11
Berries	68
Ferns	79
Rogues	88
Shrubs & Bushes	97
Broadleaf Trees	118
Conifers	140
Glossary	160
Bibliography	162
Index	163

A GREAT NATURAL GARDEN

Vancouver's extraordinary natural setting and lush climate make it one of the greatest natural gardens in the world. Mountains, forests, bogs, lakes, rivers and the ever-present ocean combine with moderate temperatures and high rainfalls to create an extremely diverse growing environment.

The Lower Mainland boasts many exceptional micro areas where plant life is quite different from areas a short distance away. A five-minute walk into Burns Bog — home to insect-eating sundews, stunted forests of shore pine, arctic starflowers, Labrador tea, bog laurel, sweet gale and gigantic skunk cabbage — makes you feel as if you have entered Jurassic Park. In contrast, the two-kilometre walk from Cypress Bowl to First Lake on Hollyburn Mountain puts you in touch with a quite different world of seemingly magical alpine plants: queen's cup, deer cabbage, mountain hemlock, twisted stalk, yellow cypress, Pacific silver fir and cow-parsnip, to name but a few. The coast mountains from Whistler through Vancouver and up the Fraser Valley are all enchanted in this fashion.

Our lakes, marshes, ponds and rivers support another diverse community of plants, including willows, poplars, silverweed, horsetails, irises, sedges, grasses, hardhack and water lilies. In some parts of the Lower Mainland, natural sites have been left untouched by chance alone; in others, individuals, parks boards and naturalist societies have worked hard to designate, preserve or restore our natural environment.

< *Early spring light in Stanley Park*

The plants described in this guide are a good representation of the plants to be found in the natural areas of Vancouver and the Lower Mainland. Each species has a fact sheet with three or four entries. DESCRIPTION presents the plant and how to identify it, while HABITAT explains where it grows best. Plants valued by First Nation's peoples have an entry for NATIVE USE. In the LOCAL SITES section, I have listed a few places in and around Vancouver where each plant can be found. Though these are some of the best areas, the lists are far from exhaustive.

Most of the place names in the LOCAL SITES entries can be found in the maps on the front and back flaps.

Please note that I have only included native plants and introduced species that survive and thrive in the wild. Thus you will search in vain for the monkey puzzle tree, a South American transplant common in Vancouver gardens but never found away from them, as it does not naturally reproduce in this region. Also note that, to keep the book pocket-sized, I have not included every known species. The observer with a keen interest and a sharp eye can expect to discover more treasures.

– Collin Varner

Early autumn sunrise in Burn's Bog

ACKNOWLEDGEMENTS

This guide could not have been prepared without the help of many individuals, organizations and government agencies. I would like to acknowledge and thank the following:

Naturalist and photographer Greg Thrift — thank you for allowing me access to your extensive photograph collection, your ability to capture nature on film is first class.

Dr. Karel Klinka for his careful proofreading.

Merry Merridith for her help with the artwork.

Brenda O'Reilly for her keyboarding skills.

The University of British Columbia.

The great cooperation of the civic parks boards, B.C. Parks and the GVRD Regional Parks Department.

Special thanks for information supplied by Duanne van den Berg, Alouette Field Naturalists; Elaine Gould, Berk Mountain Naturalists; and Daphne Solecki, Vancouver Natural History Society.

And finally, the great staff at Raincoast Books: Kevin Williams, Scott Steedman, Simone Doust, Ingrid Paulson and Marjolein Visser.

CLIMBING HONEYSUCKLE
Lonicera ciliosa • Honeysuckle family: *Caprifoliaceae*

■ **DESCRIPTION** Climbing honeysuckle is a deciduous woody vine capable of climbing trees to 8 m in height. Its orange flowers are trumpet-shaped, to 4 cm long, and form in clusters in the terminal leaves. By late summer bunches of bright red berries are produced in the cup-shaped leaves. The leaves are oval, 5-8 cm long and, like all honeysuckles, opposite. This species is the showiest of the native honeysuckles. Its main pollinators are hummingbirds and moths. CAUTION: the berries are considered poisonous.

■ **HABITAT** Scattered in low-elevation Douglas fir forests, more common near the ocean and in the Gulf Islands.

■ **NATIVE USE** The vines were used to weave mats, blankets and bags.

■ **LOCAL SITES** Can be seen climbing over bushes and trees from Caulfeild Cove to Horseshoe Bay, above the Stanley Park seawall, in Pacific Spirit Park and scattered throughout the Fraser Valley. Flowering starts at the end of May.

< *Swordfern Trail in Pacific Spirit Park.*

TWINFLOWER
Linnaea borealis • Honeysuckle family: *Caprifoliaceae*

■ DESCRIPTION Twinflower is an attractive trailing evergreen to 10 cm in height. Its nodding pink flowers are fragrant, to 5 mm long, and borne in pairs at the end of slender, Y-shaped stems. The evergreen leaves are 1 cm long, oval, shiny dark green above and paler below, with minute teeth on the upper half. The genus *Linnaea* is named for Carolus Linnaeus, Swedish botanist and founder of the binomial system for plant and animal classification. Twinflower is said to have been his favourite flower.

■ HABITAT Common in low to mid elevation forests across Canada.

■ LOCAL SITES Can be seen carpeting large areas at the top of the Grouse Grind, on Hollyburn Mountain, and in Cypress Park, Lighthouse Park and the Alouette Lake area. Flowers mid-June through July.

BULL THISTLE
Cirsium vulgare • Aster family: *Asteraceae*

■ **DESCRIPTION** Bull thistle is an introduced, well-armed biennial to 1.7 m in height. Its showy flowers are pinkish purple, well-armed towards the base, to 4 cm long. The leaves are alternate, deeply incised and armed on both the edges and the top surface. This species puts out vegetative growth in the first year and flowers in the second. The species name *vulgare* means "common."

■ **HABITAT** Disturbed sites. Because grazing animals do not eat these plants, they have spread well into fertile pastures and fields.

■ **LOCAL SITES** Occasionally seen in urban centres, but mainly on farmland in the Southlands neighbourhood of Vancouver, Boundary Bay and the Fraser Valley.

FLOWERS

WALL LETTUCE
Lactuca muralis • Aster family: *Asteraceae*

■ **DESCRIPTION** Wall lettuce is an introduced herbaceous biennial to 1.5 m tall. Its tiny yellow flowers, which resemble small dandelions, grow in loose clusters to 25 cm across. The fruit (achenes) are small and covered with fluffy hairs. Leaves are variable in size and shape, though most are deeply incised and clasp the stem. The milky sap gave the plant its genus name, *Lactuca*, from the Latin word for milk, *lac*.

■ **HABITAT** Very common in southern B.C., on roadsides and highway medians and in open forests.

■ **LOCAL SITES** Very common in all parks and rights-of-way. Flowering starts in June and sputters out in September.

OXEYE DAISY

Chrysanthemum leucanthemum • Aster family: *Asteraceae*

■ DESCRIPTION Oxeye daisy is an aromatic herbaceous perennial to 75 cm tall. Its flowers have the typical daisy white ray petals and yellow centre disks, to 5 cm across. The basal leaves are obovate with rounded teeth; the stem leaves are similar, though alternate. This is a European introduction that has naturalized on most of Vancouver Island and the Lower Mainland. "Chrysanthemum" is from the Greek *chrysos* ("gold") and *anthos* ("flower").

■ HABITAT Fields, meadows, very common on roadsides.

■ LOCAL SITES Common, with large drifts along the road up to Cypress Bowl. Often takes over abandoned fields. Flowers June through July.

FLOWERS

GUM WEED
Grindelia integrifolia • Aster family: *Asteraceae*

■ **DESCRIPTION** Gum weed is a herbaceous perennial to 1 m in height. Its yellow ray flowers grow to 5 cm across, with the bracts covered in a gummy latex. The basal leaves are yellowish green, 5-30 cm long and lance-shaped. Gum weeds are halophytes — they need salt, which they get from salt spray from the ocean.
■ **HABITAT** High beaches along the coast.
■ **NATIVE USE** The latex was used to treat asthma, bronchitis and whooping cough.
■ **LOCAL SITES** Rocky bluffs and outcrops from Caulfeild Cove to Horseshoe Bay. Flowering starts at the beginning of June.

COLTSFOOT
Petasites palmatus • Aster family: *Asteraceae*

FLOWERS

■ **DESCRIPTION** Coltsfoot is a large-leafed herbaceous perennial to 60 cm in height. Its purplish white flowers emerge before the leaves in late winter; they are grouped together to form terminal clusters approximately 10 cm across on a 60-cm stalk. The leaves have 7-9 lobes and grow to 30 cm across, green above and white and woolly below. The genus name *Petasites* is from the Greek word *petasos*, meaning "hat." Japanese children once used the large leaves as hats.

■ **HABITAT** Moist to wet areas at low to mid elevations.

■ **NATIVE USE** The leaves were used to cover berries in steam cooking pits.

■ **LOCAL SITES** Large drifts along NW Marine Drive from Locarno Beach to Wreck Beach and from outside the UBC David C. Lam Asian Garden to the Musqueam Golf Course. Flowering starts at the beginning of March. *P. frigidus* is an alpine species that can be seen in the Garibaldi-Whistler area.

COMMON TANSY
Tanacetum vulgare • Aster family: *Asteraceae*

■ **DESCRIPTION** Common tansy is an aromatic herbaceous perennial to 1.2 m in height. Its yellow, button-like flowers are grouped together to form attractive flat clusters 5-10 cm across. The delicate leaves are alternate, finely divided, 5-25 cm long and up to 10 cm wide. The entire plant has an interesting but not disagreeable odour, somewhat like camphor. Common tansy is a European introduction that has naturalized very well here. It has a long and colourful history, and is still used medicinally and as a culinary flavouring.

■ **HABITAT** Well-drained sites at low elevations.

■ **LOCAL SITES** Common in neglected areas. One of the largest sites is at Jericho Park West, where it covers a third of a hectare. A wonderful sight when in full bloom, mid-July to mid-August.

YARROW
Achillea millefolium • Aster family: *Asteraceae*

FLOWERS

■ DESCRIPTION Yarrow is a herbaceous perennial to 1 m in height. Its many small white flowers form flat-topped clusters 5-10 cm across. The aromatic leaves are so finely dissected that they appear fern-like, hence its species name "a thousand leaves." The genus is named after Achilles, a hero of Greek mythology.

■ HABITAT Roadsides, wasteland, common at low to mid elevations.

■ NATIVE USE Infusions and poultices were made for cold remedies.

■ LOCAL SITES Common throughout the Lower Mainland. Large patches seen in fields at Boundary Bay, Fraser River below Musqueam Reserve and at Minnekhada Park. Flowering starts mid-May.

CANADA GOLDENROD
Solidago canadensis • Aster family: *Asteraceae*

■ DESCRIPTION Canada goldenrod is a herbaceous perennial of various heights, from 30 to 150 cm. Its small golden flowers are densely packed to form terminal pyramidal clusters. The many small leaves grow at the base of the flowers; they are alternate, lance-linear, sharply saw-toothed to smooth.

■ HABITAT Roadsides, wasteland, forest edges at low to mid elevations.

■ LOCAL SITES Can be seen growing at the sides of roads and train tracks from Chilliwack to Whistler Village. Flowering starts mid-July and continues to autumn.

PEARLY EVERLASTING
Anaphalis margaritacea • Aster family: *Asteraceae*

■ **DESCRIPTION** Pearly everlasting grows to 80 cm in height and produces heads of small yellowish flowers surrounded by dry white bracts. The leaves are lance-shaped, green above and covered with a white felt underneath. If picked before they go to seed, the flowers remain fresh-looking long after they are brought in.

■ **HABITAT** Common on disturbed sites, roadsides and rock outcrops.

■ **LOCAL SITES** Commonly seen in waste areas. Flowering starts mid-July and flower heads can be seen into the winter.

FLOWERS

COAST BOYKINIA
Boykinia elata • Saxifrage family: *Saxifragaceae*

■ **DESCRIPTION** Coast boykinia is a herbaceous perennial to 60 cm in height. Its small white flowers are produced on long slender stalks 25-60 cm tall. The grass green leaves are somewhat heart-shaped with 5-7 lobes, to 8 cm across, and supported on long slender hairy stems. When in flower, boykinia is very attractive. The species name *elata* means "tall."

■ **HABITAT** Moist forests, wet cliff faces and streamsides at low to mid elevations.

■ **LOCAL SITES** Wet shaded cliffs above the Stanley Park seawall, lower levels of North and West Vancouver mountains and in Lighthouse Park. Flowers June and July.

SMALL-FLOWERED ALUMROOT
Heuchera micrantha • Saxifrage family: *Saxifragaceae*

■ **DESCRIPTION** Small-flowered alumroot is a perennial to 60 cm in height. Its small white flowers are abundant and held on scapes (stems) up to 60 cm tall. The heart-shaped leaves have long hairy stems and are basal. The leaves are slightly longer than they are broad and distinguish this plant from smooth alumroot (*H. glabra*), which has leaves that are broader than they are long. The name "alumroot" is given because the roots are very astringent.

■ **HABITAT** Wet cliff faces and stream banks at low to high elevations.

■ **LOCAL SITES** Wet cliffs on the North Shore mountains to Shannon Falls. Can be seen on cliffs above the Stanley Park seawall and at Lighthouse Park, and masses blanket the rock faces on the Sea to Sky Highway (Hwy 99) between Horseshoe Bay and Lions Bay. Flowering starts mid-May.

FLOWERS

FRINGECUP
Tellima grandiflora • Saxifrage family: *Saxifragaceae*

■ **DESCRIPTION** Fringecup is a perennial to 80 cm in height. Its fringed flowers are greenish, fragrant, 1 cm long and produced on 60- to 80-cm scapes (stems). The basal leaves are round to heart-shaped, deeply notched, 5-8 cm across; the scape leaves are smaller. When out of flower, fringe-cup can be confused with the piggy-back plant (*Tolmiea menziesii*; see opposite).

■ **HABITAT** Moist cool forests along the coast.

■ **NATIVE USE** The plants were crushed and boiled and the resultant infusion was used to treat sickness.

■ **LOCAL SITES** Common in forested areas such as Pacific Spirit Park, Richmond Nature Park and Deer Lake, usually near shaded cliffs or stream banks. Flowering starts end of April and continues to June.

PIGGY-BACK PLANT
Tolmiea menziesii • Saxifrage family: *Saxifragaceae*

■ **DESCRIPTION** The piggy-back gets its unusual name from the way it reproduces. By late summer young plants can be seen growing at the base of the leaves. As autumn approaches the leaves fall to the ground, allowing the young plants to take root. Mature plants can grow to a height of 70 cm. The flowers are small, reddish purple and inconspicuous. The dark green leaves are rough, with 5-7 lobes and up to 10 cm across.

■ **HABITAT** Moist forests at low to mid elevations.

■ **LOCAL SITES** A common plant in forests, but it blends well with other plants and is often overlooked. Flowering starts mid-May.

FOAM FLOWER
Tiarella trifoliata • Saxifrage family: *Saxifragaceae*

■ DESCRIPTION Foam flower is a herbaceous perennial to 50 cm in height. Each wiry stem supports several tiny white flowers. The massed flowers are thought to resemble foam. The trifoliate leaves (to 7 cm across) are all basal except for one, located approximately halfway up the stem; this is good for identification. There is another species of foam flower (*T. unifoliata*) that is very similar except for its solid leaf.

■ HABITAT Shaded moist woods at low to mid elevations.

■ LOCAL SITES Very common in forests, such as Central Park in Burnaby and Golden Ears Park. Can often be seen growing in masses. Flowering starts mid-May and continues through July. The solid-leaf foam flower is less common and can be seen in the Whistler area.

HARVEST LILY
Brodiaea coronaria • Lily family: *Liliaceae*

FLOWERS

■ DESCRIPTION Harvest lily is a herbaceous perennial, to 30 cm in height from corms. Its violet-purple, trumpet-shaped flowers are 4 cm long and grow in clusters of 3-5. The leaves are grass-like and have withered by the time the flowers are noticeable.

■ HABITAT Isolated to southeastern Vancouver Island, the Gulf Islands and adjacent mainland. Prefers well-drained grassy slopes.

■ NATIVE USE The corms were harvested for winter consumption.

■ LOCAL SITES Caulfeild Cove, Lighthouse Park and Whytecliff Park. Fool's onion (*B. hyacinthina*; see page 30) is also found in this area. Not common. Flowers mid-July.

FLOWERS

TWISTED STALK
Streptopus amplexifolius • Lily family: *Liliaceae*

■ **DESCRIPTION** Twisted stalk is a branching, herbaceous perennial, 1-2 m in height. Its greenish white flowers are 1 cm long and borne in leaf axils on slender twisted stalks. The fruit develops into a bright red oval berry to 1 cm long. The ovate leaves are alternate, 5-12 cm long. They clasp the stem directly, with no petiole; the species name *amplexifolius* means "clasping leaves." The flowers and fruit hang from the leaf axils along the branches.
CAUTION: the berries are considered poisonous.
■ **HABITAT** Cool moist forests at low to high elevations.
■ **NATIVE USE** The plants were tied to the clothing or hair for their scent.
■ **LOCAL SITES** Cypress Bowl, Hollyburn Mountain, Lighthouse Park to Harrison Lake at Bear Mountain. Flowering starts at the beginning of June in low areas, in late June at higher elevations.

QUEEN'S CUP or BEAD LILY
Clintonia uniflora • Lily family: *Liliaceae*

■ DESCRIPTION Queen's cup is a herbaceous perennial from 8 to 15 cm in height. Its white, cup-shaped flowers grow to 3 cm across, usually with only one borne at the end of a slender stalk. The cobalt blue fruit is round to pear-shaped and singular. The 2 to 5 bright green leaves are broadly lance-shaped, basal and fleshy. The genus name commemorates DeWitt Clinton, governor of New York State, botanist and developer of the Erie Canal.

■ HABITAT Cool moist coniferous forests at low to high elevations.

■ LOCAL SITES Huge patches form a ground cover on the pathway from Cypress Bowl parking lot to First Lake on Hollyburn Mountain and the pathway to Brandywine Falls. Flowering starts end of June, with berries ripening mid-August.

FOOL'S ONION or WILD HYACINTH
Brodiaea hyacinthina • Lily family: *Liliaceae*

■ **DESCRIPTION** Fool's onion is a herbaceous perennial, to 60 cm in height from corms. The small star-shaped flowers are white with fine stripes of green, held up in terminal clusters on thin stems up to 60 cm long. The grass-like leaves are basal and have usually disappeared by the time the flowers are noticeable.

■ **HABITAT** Rocky outcrops, grassy slopes, mainly on the southeastern side of Vancouver Island and adjacent mainland.

■ **NATIVE USE** The corms were collected and eaten raw or boiled.

■ **LOCAL SITES** A few select pockets around Caulfeild Cove, and at Lighthouse Park and Whytecliff Park. Flowers at the beginning of June. Harvest lily (*B. coronaria*; see page 27) is also found in this area.

INDIAN HELLEBORE or CORN LILY
Veratrum viride • Lily family: *Liliaceae*

FLOWERS

■ **DESCRIPTION** Indian hellebore is a tall, herbaceous perennial from 1 to 2 m in height. Its strongly ribbed, grass green leaves are ovate to elliptic, 10-30 cm long, with a passing resemblance to corn leaves. The numerous yellowish green flowers hang in drooping panicles from the upper portion of the plant.
CAUTION: this plant is extremely poisonous.

■ **HABITAT** Moist open forests at low to high elevations and wet alpine meadows and swales.

■ **NATIVE USE** Indian hellebore was known for its magical power and highly valued by virtually all coastal people. Although it is poisonous, it was used as a medicine and to ward off evil spirits. It was known as a *skookum* ("strong") medicine.

■ **LOCAL SITES** Top of Grouse Grind, and at Cypress Bowl and First Lake on Hollyburn Mountain. Flowers from June to August, depending on altitude.

FLOWERS

NODDING ONION
Allium cernuum • Lily family: *Liliaceae*

■ **DESCRIPTION** Nodding onion is a herbaceous perennial, to 45 cm in height from bulb. Over a dozen small pink flowers are held in the distinctive nodding umbels. The grassy leaves are basal, to 30 cm long, and similar to those of a green onion. Both bulbs and leaves smell of onion. The species name *cernuum* means "nodding."

■ **HABITAT** Dry grassy slopes, rocky outcrops, forest edges at lower elevations.

■ **NATIVE USE** The cooked onions were a delicacy. In the Salish language, *lillooet* means "place of many onions."

■ **LOCAL SITES** Scattered in rocky crevices from Caulfeild Cove to Whytecliff Park. Flowering starts at the beginning of June.

DEATH CAMAS
Zygadenus venenosus • Lily family: *Liliaceae*

■ **DESCRIPTION** Death camas is a herbaceous perennial, to 50 cm in height from bulb. Its small creamy flowers are 1 cm across and neatly arranged in terminal racemes on stems to 50 cm long. The grass-like leaves are mainly basal, to 30 cm long, and have a deep groove like a keel down the centre. The entire plant is poisonous and when out of flower can be confused with the edible common camas (*Camassia quamash*).
CAUTION: the entire plant is poisonous.

■ **HABITAT** Rocky outcrops and grassy slopes at low elevations.

■ **NATIVE USE** The bulbs were mashed and used as arrow poison.

■ **LOCAL SITES** On rocky outcrops from Caulfeild Cove to Whytecliff Park. Seen more on the Gulf Islands and in the B.C. Interior. Flowers from late April through May.

BLACK LILY or NORTHERN RICE ROOT
Fritillaria camschatcensis • Lily family: *Liliaceae*

■ **DESCRIPTION** Black lily is a strong herbaceous perennial to 60 cm in height. Its nodding flowers have 6 petals and are bell-shaped, purple-brown, to 3 cm across. The leaves are in whorls, lance-shaped, to 8 cm long. The alternative name "rice root" comes from the large white bulbs, which are covered with rice-like scales. Chocolate lily (*F. lanceolata*) is a less sturdy species with thinner leaves and mottled flowers. The genus name *Fritillaria* refers to the flower's checkered pattern, reminiscent of old dice boxes.

■ **HABITAT** Open forests and moist grassy fields at low to high elevations.

■ **NATIVE USE** The bulbs were boiled or steamed and eaten by most Pacific Northwest peoples.

■ **LOCAL SITES** Both species grow at lower elevations from Caulfeild Cove to Horseshoe Bay. Black lily can be seen in the damp grassy areas at Noons Creek in Port Moody (please stay on the boardwalk). Flowering starts mid-May.

FALSE LILY OF THE VALLEY
Maianthemum dilatatum • Lily family: *Liliaceae*

■ **DESCRIPTION** False lily of the valley is a small herbaceous perennial to 30 cm in height. Its small white flowers appear in April/May, clustered on 5-10 cm spikes. The slightly fragrant flowers are quickly replaced by berries 6 mm across; the berries go through summer a speckled green and brown but turn ruby red by autumn. The dark green leaves are alternate, heart-shaped and slightly twisted, to 10 cm long. The genus name *Maianthemum* is from the Greek *Maios* ("May") and *anthemon* ("blossom").

■ **HABITAT** Moist coastal forests at low elevations.

■ **NATIVE USE** The berries were eaten but not highly regarded.

■ **LOCAL SITES** Common in forested areas such as around Mundy Lake, Noons Creek in Port Moody and Deer Lake, and in Central Park, Stanley Park and Pacific Spirit Park. The leaves emerge mid-March and form carpets by the time the flowers bloom in early May. Berries start showing mid-June.

TIGER LILY
Lilium columbianum • Lily family: *Liliaceae*

- **DESCRIPTION** Tiger lily is an elegant herbaceous perennial to 1.5 m tall. Its drooping flowers go from deep yellow to bright orange. A vigorous plant can have 20 or more flowers. Shortly after the flower buds have opened, the tepals curve backwards to reveal maroon spots and anthers. The leaves are lance-shaped, usually in a whorl and 5-10 cm long. It is said that he or she who smells a tiger lily will develop freckles.
- **HABITAT** Diverse range, including open forests, meadows, rock outcrops and the sides of logging roads, at low to sub-alpine elevations.
- **NATIVE USE** The bulbs were boiled or steamed and eaten.
- **LOCAL SITES** Caulfeild Cove, Lighthouse Park, Golden Ears Park and Harrison Lake. Flowering starts mid-May at lower elevations to August at sub-alpine.

WESTERN WHITE FAWN LILY
Erythronium oregonum • Lily family: *Liliaceae*

■ **DESCRIPTION** Western white fawn lily is a herbaceous perennial to 25 cm in height. The nodding white flowers are adorned with golden anthers, and the seed takes 5 to 7 years to form a corm and put up its first flower; picking of the flowers has greatly reduced the numbers of this plant. The basal leaves are lance-shaped to 20 cm long and mottled white to brown, much like a fawn.

■ **HABITAT** Open forests and rocky outcrops at low elevations.

■ **LOCAL SITES** Alouette Lake. Good displays can also be seen at Lighthouse Park towards the end of March.

WESTERN TRILLIUM
Trillium ovatum • Lily family: *Liliaceae*

■ **DESCRIPTION** The western trillium's beautiful flowers are made up of 3 white petals, approximately 5 cm long, and 3 green sepals, elevated on a stem 30-50 cm tall. The flowers change from white to pink to a mottled purple before withering away. The dark green leaves, usually arranged in whorls of threes, are widely ovate and up to 15 cm long. Development, logging and overpicking have led to a dramatic decrease in wild trilliums. They are now protected by law.

■ **HABITAT** Moist forested areas in southern B.C. at lower elevations.

■ **NATIVE USE** The fleshy rhizomes were used for medicinal purposes.

■ **LOCAL SITES** Mundy Park has one of the best showings, and there are hidden pockets at Harrison Lake, Alouette Lake, Grouse Mountain, Hollyburn/Cypress, Lighthouse Park and Pacific Spirit Park. Flowering starts at the beginning of April.

LARGE-LEAFED AVENS
Geum macrophyllum • Rose family: *Rosaceae*

■ **DESCRIPTION** Large-leafed avens is a herbaceous perennial to 90 cm in height. Its bright yellow flowers resemble buttercups. They are approximately 6 mm across and are produced singularly or in small clusters. The unique round fruit has bristly bent protruding styles that catch on fur and clothing, an excellent way of dispersing the seed. The irregular-shaped larger leaves are 15-20 cm across, while the stem leaves are smaller and 3-lobed.

■ **HABITAT** Prefers moist soil in open forests and beside pathways, trails and roads at low elevations.

■ **NATIVE USE** The roots were boiled and used medicinally.

■ **LOCAL SITES** Common throughout the Lower Mainland, such as in Stanley Park, Central Park and at Harrison Lake; often found in association with fringecup (*Tellima grandiflora*). Flower starts at the end of April and continues irregularly through August.

FLOWERS

SILVERWEED
Potentilla anserina ssp. *pacifica* • Rose family: *Rosaceae*

■ **DESCRIPTION** Silverweed only grows to 30 cm in height but can take over several hectares in favourable conditions. The yellow flowers are produced singly on a leafless stalk. The compound leaves reach 25 cm in length and have 9-19 toothed leaflets; they are bicoloured, grass green above and felty silver below, hence the common name. Silverweed spreads quickly thanks to its fast-growing stolons, which root at the nodes. The genus name *Potentilla* means "powerful," a reference to its medicinal properties.
■ **HABITAT** Saline marshes, meadows and wet run-off areas near the ocean.
■ **NATIVE USE** The cooked roots were an important food source.
■ **LOCAL SITES** Common in the marshy areas of the Fraser Valley and at the bottom of Boom Trail in Pacific Spirit Park, Jericho Beach Park pond, Noons Creek in Port Moody and Lighthouse Park. Flowering starts mid-May.

YELLOW MONKEY-FLOWER
Mimulus guttatus • Figwort family: *Scrophulariaceae*

■ DESCRIPTION Yellow monkey-flower can be annual or perennial; normally it self-seeds, disappears and then germinates as an annual in spring. It can grow to 80 cm in height. Its beautiful yellow flowers are 2-lipped, to 5 cm long, with many small and one larger dot on the lower lip. The lower leaves are oval and grow in pairs, while the upper leaves hug the stem. Chickweed monkey-flower (*M. alsinoides*) is smaller, to 20 cm high, and often grows with the larger variety; its flowers are much smaller and have only a single dot on the lower lip.
■ HABITAT Wet cliffs and ledges at low elevations.
■ LOCAL SITES The cliffs above Wreck Beach and between Dunbar and Trafalgar beaches have sheets of yellow flowers from mid-May to June. Flowering sputters on through July and August.

FOXGLOVE
Digitalis purpurea • Figwort family: *Scrophulariaceae*

■ DESCRIPTION Foxglove is a biennial from 1 to 2 m in height. Its tall spikes of flowers appear in the second year after germination. The drooping flowers are 5 cm long, pink to purple, with darker spots on the inside. The basal leaves are soft, hairy and large — 15-30 cm long — but decrease in size towards the top of the plant. Foxglove is an introduced plant from Europe that has become widespread throughout urban B.C. The heart drug digitalis is made from it.

CAUTION: the entire plant is very poisonous.

■ HABITAT Moist disturbed sites at lower elevations.

■ LOCAL SITES Waste areas, beach cliffs, roadsides, fields and, of course, gardens.

COW-PARSNIP or INDIAN CELERY
Heracleum lanatum • Carrot family: *Apiaceae*

■ **DESCRIPTION** Cow-parsnip is a tall hollow-stemmed herbaceous perennial from 1 to 3 m in height. Its small white flowers are grouped in flat-topped umbrella-like terminal clusters to 25 cm across. It produces numerous small, egg-shaped seeds, 1 cm long, with a pleasant aroma. The large woolly compound leaves are divided into 3 leaflets, 1 terminal and 2 lateral (to 30 cm across). The genus name *Heracleum* is fitting for this plant of Herculean proportions. Giant cow-parsnip (*H. mantegazzianum*), an introduced species, grows to 4 m in height and can be seen in urban areas.

CAUTION: both species can cause severe blistering and rashes when handled.

■ **HABITAT** Moist forests, meadows, marshes and roadsides from low to high elevations.

■ **LOCAL SITES** Damp forested slopes. Giant cow-parsnip can be seen growing beside NW Marine Drive, at the north end of Lions Gate Bridge, below Prospect Point in Stanley Park, on the slopes west of the Vancouver Maritime Museum and along the pathway between Trafalgar and Kitsilano Beach. Plants emerge early March, flowering starts end of May.

WILD CARROT or QUEEN ANNE'S LACE
Daucus carota • Carrot family: *Apiaceae*

■ **DESCRIPTION** Wild carrot is an introduced biennial to 1 m in height. Its small white flowers are grouped together to form showy terminal clusters to 10 cm across. The leaves (to 15 cm long) are dissected to the point that they resemble delicate ferns. If the stems are scratched, a carrot scent is released. Wild carrot and parsnips have long been cultivated in Europe for culinary and medicinal use; records of them date back to 500 B.C.

■ **HABITAT** Roadsides, abandoned fields and highway medians at low elevations.

■ **LOCAL SITES** Common from Horseshoe Bay to Harrison Lake. Flowers July through September.

WATER-PARSLEY
Oenanthe sarmentosa • Carrot family: *Apiaceae*

■ **DESCRIPTION** Water-parsley is a semi-aquatic herbaceous perennial to 1 m in height. Its flowers are white, faintly fragrant and borne in flat-topped clusters. The leaves are pinnately divided 2 or 3 times, with deeply toothed leaflets. The overall appearance of the plant is weak and sprawling.

CAUTION: the entire plant is considered poisonous.

■ **HABITAT** Low-elevation marshes and swamps, occasionally in ditches.

■ **LOCAL SITES** Common in wet areas throughout the Lower Mainland, with thick concentrations along the banks of the Fraser River in the Southlands neighbourhood of Vancouver. Flowers mid-June through July.

COOLEY'S HEDGE-NETTLE
Stachys cooleyae • Mint family: *Lamiaceae*

■ **DESCRIPTION** Cooley's hedge-nettle is a herbaceous perennial to 1 m in height. Its purply red flowers are trumpet-like with a lower lip; they grow to 4 cm long and are grouped in terminal clusters. The leaves are mint-like with toothed edges, opposite, finely hairy on both sides, to 15 cm long. The stems are square and finely hairy. Cooley's hedge-nettle was first documented in 1891 by Grace Cooley, a professor from New Jersey who saw it near Nanaimo.

■ **HABITAT** Moist open forests and streamsides at low elevations.

■ **LOCAL SITES** Common in moist fields and lower forest edges. Flowers in June to mid-July.

HEAL-ALL or SELF-HEAL
Prunella vulgaris • Mint family: *Lamiaceae*

■ **DESCRIPTION** Heal-all is an introduced herbaceous perennial to 40 cm in height. Its purple flowers are 2-lipped, 1-2 cm long, and borne in terminal spikes. The leaves are mostly lance-shaped, opposite, to 7 cm long. The stems are square. As its name suggests, heal-all has long been used medicinally. Seventeenth-century herbalist Nicholas Culpeper prescribed that it be "taken inwardly in syrups for inward wounds, outwardly in unguents and plasters for outward."

■ **HABITAT** Roadsides, forest edges, fields and parks at low elevations.

■ **LOCAL SITES** Common on naturalized grassed areas in most parks and forest edges. Flowers through most of the summer.

FLOWERS

CREEPING BUTTERCUP
Ranunculus repens • Buttercup family: *Ranunculaceae*

■ **DESCRIPTION** Creeping buttercup is a prostrate creeping perennial that rarely grows above 30 cm in height. The flowers are 2-3 cm across; their solid deep yellow colour contrasts well with the dark green foliage. The pale blotched leaves have long stalks and are divided into 3 leaflets.
CAUTION: they may not look threatening, but buttercups are considered poisonous and can cause skin irritations.
■ **HABITAT** Moist to wet sites in urban areas, parks, fields, and open forests.
■ **LOCAL SITES** Commonly seen in lawns, parks and fields. Starts flowering at the beginning of May. Several other taller species can also be seen.

RED COLUMBINE
Aquilegia formosa • Buttercup family: *Ranunculaceae*

■ DESCRIPTION Red columbine is a herbaceous perennial to 1 m in height. The drooping red-and-yellow flowers are up to 5 cm across and have 5 scarlet spurs arching backwards; they are almost translucent when the sun shines on them. The leaves are sea green above, paler below, to 8 cm across and twice divided by threes. In the head of the flower is a honey gland that can only be reached by hummingbirds and long-tongued butterflies. The hole that can sometimes be seen above this gland is caused by frustrated bumblebees chewing their way to the nectar. The name "columbine" means "dove," for the five arching spurs said to resemble five doves sitting around a dish.

■ HABITAT Moist open forests, meadows and creeksides at low to high elevations.

■ LOCAL SITES Bowen Island, Golden Ears Park and higher elevations to Whistler Mountain, where it forms small patches. Flowers mid-June at lower and mid-August at higher elevations.

PACIFIC BLEEDING HEART
Dicentra formosa • Bleeding heart family: *Fumariaceae*

■ **DESCRIPTION** B.C.'s native bleeding heart is very familiar, thanks to its resemblance to the many cultivated varieties. It is a herbaceous perennial to 40 cm in height, with pinkish heart-shaped flowers that hang in clusters of 5-15. The delicate fern-like leaves are basal, but sweep upwards so much that they almost hide the flowers. Under good growing conditions, bleeding hearts can cover hundreds of square metres. The species name *formosa* means "beautiful."

■ **HABITAT** Open broad-leafed forests with nutrient-rich topsoil.

■ **LOCAL SITES** Commonly seen in lower forested areas; grows as ground cover in the northern portion of Pacific Spirit Park. Flowering starts mid-April and lasts to the end of May, with the entire plant collapsing by the beginning of July.

WESTERN STAR FLOWER
Trientalis latifolia • Primrose family: *Primulaceae*

■ **DESCRIPTION** Western star flower is a small, herbaceous perennial 10-25 cm in height. Its white to pink flowers hang on very thin stalks, making them appear like stars. The oval leaves (5-10 cm long) are elevated in a whorl just under the flower stalks. There is a northern star flower (*T. arctica*) that is more confined to bogs and swamps; it is shorter (5-20 cm in height), with white flowers 1.5 cm across and additional leaves on the stem below the whorl of elevated leaves.
■ **HABITAT** Dry to moist coniferous forests at low elevations.
■ **LOCAL SITES** Common in Lower Mainland forests. Flowering starts mid-May and continues to the end of June. Northern star flower can be seen at Camosun Bog and Burns Bog. Flowering starts in June.

FLOWERS

FIREWEED
Epilobium angustifolium•
Evening primrose family: *Onagraceae*

■ DESCRIPTION Fireweed is a tall herbaceous perennial that reaches heights of 3 m in good soil. Its purply red flowers grow on long showy terminal clusters. The leaves are alternate, lance-shaped like a willow's, 10-20 cm long and darker green above than below. The minute seeds are produced in pods 5-10 cm long and have silky hairs for easy wind dispersal. Fireweed flowers have long been a beekeeper's favourite. The name "fireweed" comes from the fact that it is one of the first plants to grow on burned sites; typically follows wildfires.

■ HABITAT Common throughout B.C. in open areas and burned sites.

■ NATIVE USE The stem fibres were twisted into twine and made into fishing nets, and the fluffy seeds were used in padding and weaving.

■ LOCAL SITES A common roadside plant. Plants emerge mid-April and flower from mid-June to August.

BUNCHBERRY or DWARF DOGWOOD
Cornus canadensis • Dogwood family: *Cornaceae*

FLOWERS

■ DESCRIPTION Bunchberry, a perennial no higher than 20 cm, is a reduced version of the Pacific dogwood tree (*C. nuttallii*). The tiny greenish flowers are surrounded by 4 showy white bracts, just like the flowers of the larger dogwood. The evergreen leaves, 4-7 cm long, grow in whorls of 5-7 and have parallel veins like the larger tree's. The beautiful red berries form in bunches (hence the name) just above the leaves in August. Bunchberry and Pacific dogwood have a habit of flowering twice, once in spring and again in late summer.

■ HABITAT From low to high elevations in cool moist coniferous forests and bogs.

■ LOCAL SITES Common in forested areas of the Lower Mainland, across Grouse Mountain and Hollyburn Mountain to Shannon Falls. Large concentrations at Noons Creek in Port Moody, Deer Lake and Burns Bog.

SKUNK CABBAGE
Lysichiton americanum • Arum family: *Araceae*

- **DESCRIPTION** Skunk cabbage is a herbaceous perennial to 1.5 m in height and as much as 2 m across. The small greenish flowers are densely packed on a fleshy spike and surrounded by a showy yellow spathe, the emergence of which is a sure sign that spring is near. The tropical-looking leaves can be over 1 m long and 50 cm wide.
- **HABITAT** Common at low elevations in wet areas such as springs, swamps, seepage areas and floodplains.
- **NATIVE USE** Skunk cabbage roots were cooked and eaten in spring in times of famine. It is said this poorly named plant has saved the lives of thousands.
- **LOCAL SITES** Common in swampy forested areas such as Beaver Lake, Burns Bog and Mundy Lake. Flowering starts mid-March.

FLOWERS

GIANT KNOTWEED
Polygonum sachalinense • Buckwheat family: *Polygonaceae*

■ **DESCRIPTION** Giant knotweed is a tall herbaceous perennial to 4 m in height. Its tiny greenish white flowers hang in panicles to 13 cm long. The grass green leaves are heart-shaped and grow to 20 cm long and 12 cm across. The huge stems are bamboo-like and as much as 3 cm thick; they persist through the winter before collapsing in spring. Giant knotweed is an introduction from Japan that has naturalized in most settled areas of southern B.C.
■ **HABITAT** Prefers moist loose soil, where its invasive rhizomes can extend to form extensive patches. Lower elevations.
■ **LOCAL SITES** Ditches, roadsides and wasteland from Chilliwack to Squamish.

EVERGREEN VIOLET
Viola sempervirens • Violet family: *Violaceae*

■ DESCRIPTION Evergreen violet is a small creeping perennial to 8 cm in height. Its yellow flowers are solitary, 1-5 cm across, with delicate brown veins on the bottom petals. The evergreen leaves grow to 3 cm across and are broadly heart-shaped and leathery. The smallest of B.C.'s yellow violets, it spreads by sending out slender trailing stems. It is said that violets worn around the head will dispel the fumes of wine and prevent headache and dizziness.

■ HABITAT Dry to moist forests at low to mid elevations.

■ LOCAL SITES Common in forested areas, often hiding under bushes or fallen leaves; large patches at Alouette Lake, Pacific Spirit Park and Lighthouse Park. Flowering starts mid-March. The yellow wood violet (*V. glabella*) is abundant around Alouette Lake.

SIBERIAN MINER'S LETTUCE
Claytonia sibirica • Purslane family: *Portulacaceae*

■ **DESCRIPTION** Siberian miner's lettuce is a small annual to 30 cm in height. Its small white to pink flowers are 5-petalled and produced in abundance on long, thin, fleshy stems. The basal leaves are long-stemmed, opposite, ovate and, like the stems, succulent. Another species, *C. perfoliata*, differs in that its upper leaves are disc-shaped and fused to other flower stems. Siberian miner's lettuce was first discovered in Russia, where it was a staple food for miners. Early prospectors and settlers found both species made excellent early-season salad greens.
■ **HABITAT** Moist forest areas at low to mid elevations.
■ **LOCAL SITES** Both species are common in moist forested areas, such as Pacific Spirit Park, Lynn Headwaters Regional Park and Mount Seymour Park. Flowering starts mid-April and continues through July.

RATTLESNAKE PLANTAIN
Goodyera oblongifolia • Orchid family: *Orchidaceae*

■ **DESCRIPTION** Rattlesnake plantain is an evergreen perennial to 40 cm in height. Its numerous small flowers are greenish white, orchid-shaped and produced on a spike 20-40 cm high; they have a tendency to grow on one side of the spike. The evergreen leaves are basal and rosette-like, from 5-10 cm long. They are criss-crossed by whitish veins, creating the rattlesnake pattern that gives the plant its common name.

■ **HABITAT** Usually found in dry to moist coniferous forests with a moss-dominated understory.

■ **LOCAL SITES** Common on dry mossy slopes from Whistler to Harrison Lake. Lighthouse Park has good concentrations. Flowering starts at the end of July.

FLOWERS

HERB ROBERT
Geranium robertianum • Geranium family: *Geraniaceae*

■ **DESCRIPTION** Herb robert is an introduced, rapid-spreading annual to 45 cm in height. Its flowers (1.5 cm across) range from light pink to reddy purple. The leaves are divided into 3-5 sections and these are divided again. The fruit are pointed capsules to 2 cm long; *geranium* comes from the Greek *geranos* ("crane"), a reference to the beak-like fruit.

■ **HABITAT** Prefers moist, lightly shaded areas at low elevations.

■ **LOCAL SITES** Quite common in the Lower Mainland, mainly in urban areas such as Deer Lake, Burnaby Mountain Park and Whytecliff Park. Flowering starts in March and continues until the first frost.

VANILLA LEAF
Achlys triphylla • Barberry family: *Berberidaceae*

FLOWERS

■ **DESCRIPTION** Vanilla leaf is a herbaceous perennial to 30 cm in height. Its small white flowers are formed on a spike that stands above the leaf. The small fruit (achenes) is crescent-shaped and greenish to reddish purple. The wavy leaves have long stems and are divided into 3 leaflets, one at each side and the third at the tip. When dried, the leaves have a faint vanilla-like scent.
■ **HABITAT** Dry to moist forests at low to mid elevations in southern B.C.
■ **NATIVE USE** The leaves were used as an insect repellent.
■ **LOCAL SITES** Large patches in the Harrison Lake area and Golden Ears Park. Flowers June to July.

FLOWERS

YELLOW FLAG
Iris pseudacorus • Iris family: *Iridaceae*

■ **DESCRIPTION** Yellow flag is a semi-aquatic, herbaceous perennial to 1.2 m in height. Its showy flowers are bright yellow with dark pencilled veins on the lower lips. Each flowering stem usually bears 2 flowers. The leaves are wider (to 4 cm) and taller (to 1.2 m) than most irises. The plant is named after the rainbow goddess Iris for its diversity of flower colours. It is also thought to be the flower symbolized by the fleur-de-lys.

■ **HABITAT** Prefers shallow fresh water, in ditches, ponds and on lakeshores. A European introduction, it has become invasive in some wetland areas.

■ **LOCAL SITES** Commonly seen in ponds and lakes; huge patches at Jericho Beach Park pond and the marshes along the Fraser River at the bottom of Boom Trail in Pacific Spirit Park. Flowers mid-May through June.

PURPLE LOOSESTRIFE
Lythrum salicaria • Loosestrife family: *Lythraceae*

■ DESCRIPTION Purple loosestrife is an introduced herbaceous perennial to 1.8 m in height. Its flowers are reddish purple to pink and arranged in whorls on terminal spikes. The leaves are lance-shaped to 15 cm long, mostly opposite but sometimes in whorls clasping the stem. The genus name *Lythrum* is from the Greek *lythron* ("blood"), a reference to the colour of the flowers. *Salicaria* means "willow-like," referring to the leaves.

■ HABITAT Ponds, ditches and lakesides at lower elevations. It has become an invasive wetland weed that is overtaking many indigenous plants.

■ LOCAL SITES Campaigns to eliminate this plant have drastically reduced its numbers, but it can still be seen in most Lower Mainland ponds and lakes. Flowering begins mid-July.

FLOWERS

EUROPEAN BITTERSWEET
Solanum dulcamara • Potato family: *Solanaceae*

■ **DESCRIPTION** Bittersweet is an introduced perennial shrub-vine to 3 m in length. Its unique bluish purple flowers are folded back and have projecting yellow stamens. The berries start off green, then turn orange and finally brilliant red. The upper leaves have small lobes at their base, to 8 cm long. The name "bittersweet" refers to the root and stem, which taste first bitter and then sweet when chewed.

CAUTION: the leaves and berries are considered poisonous.

■ **HABITAT** Parks, beaches and laneways, usually in moist rich soils close to urban areas.

■ **LOCAL SITES** Common in unattended areas such as in the wooded patches on the banks of the Fraser River and at Jericho Beach Park and Trafalgar and Kitsilano beaches.

Shannon Falls near Squamish >

BLACKBERRIES
Rose family • Rosaceae

■DESCRIPTION Of B.C.'s three blackberry species, only one is native to the region. The two introduced species require more sunshine to thrive. The three are easy to identify:
Trailing blackberry (*Rubus ursinus*) — the first to bloom (end of April) and set fruit (mid-July), it is often seen rambling over plants in and out of forested areas. The berries are delicious and the leaves can be steeped as a tea.
Himalayan blackberry (*R. discolor*) — this blackberry was introduced from India and has now taken over much of the Pacific Northwest. It is heavily armed, grows rampant to 10 m and is a prolific producer of berries. Blooming starts mid-June and the fruit sets by mid-August.
Cutleaf blackberry (*R. laciniatus*) — introduced from Europe, this berry is very similar to the Himalayan blackberry but less common.
■HABITAT Common on open wasteland, forest edges, roadsides and in ditches.

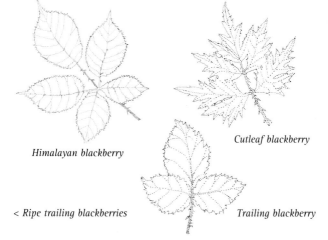

Himalayan blackberry

Cutleaf blackberry

< *Ripe trailing blackberries*

Trailing blackberry

SASKATOON BERRY or SERVICEBERRY
Amelanchier alnifolia • Rose family: *Rosaceae*

■ **DESCRIPTION** Depending on growing conditions, the saskatoon berry can vary from a 1-m shrub to a small tree 7 m in height. The white showy flowers range from 1 to 3 cm across and often hang in pendulous clusters. The young reddish berries form early and by midsummer darken to a purple black up to 1 cm across. The light bluish green leaves are deciduous, oval-shaped and toothed above the middle.

■ **HABITAT** Shorelines, rocky outcrops and open forests at low to mid elevations.

■ **NATIVE USE** The berries were eaten fresh, mixed with other berries or dried for future use. On the great plains the berries were mashed with buffalo meat to make pemmican. The hard straight wood was a favourite for making arrows.

■ **LOCAL SITES** Lower rocky bluffs around Caulfeild Cove, Lighthouse Park, Whytecliff Park and Horseshoe Bay; lots above the Stanley Park seawall. Flowering starts mid-April and the berries are fully ripe by the first week of August.

THIMBLEBERRY
Rubus parviflorus • Rose family: *Rosaceae*

■ DESCRIPTION Thimbleberry is an unarmed shrub to 3 m in height. Its large white flowers open up to 5 cm across and are replaced by juicy bright red berries. The dome-shaped berries are 2 cm across and bear little resemblance to a thimble. The maple-shaped leaves grow up to 25 cm across and, when needed, make a good tissue substitute.

■ HABITAT Common in coastal and interior B.C. in open forests at low to mid elevations.

■ NATIVE USE The large leaves were used to line cooking pits and cover baskets. The berries were eaten fresh, dried or mixed with other berries.

■ LOCAL SITES Common in forested areas, such as Shannon Falls, Stanley Park and Campbell Valley Regional Park. Flowering starts mid-May and the fruit matures at the end of July or early August.

OVAL-LEAFED BLUEBERRY
Vaccinium ovalifolium • Heather family: *Ericaceae*

■DESCRIPTION The oval-leafed blueberry is one of B.C.'s most recognized and harvested blueberries. It is a mid-size bush to 2 m in height. The pinkish bell-shaped flowers appear before the leaves and are followed by the classic blueberries. Rubbing the berries reveals a covering of dull bloom and a darker berry. The soft green leaves are smooth-edged, alternate, egg-shaped (no point) and grow to 4 cm in length.

■HABITAT Moist coniferous forests from sea level to high elevations.

■NATIVE USE Blueberries were a valuable and delicious food source. They were eaten fresh, mixed with other berries and dried for future use. As with all blueberries, they were also mashed to create a purple dye used to colour basket materials.

■LOCAL SITES Common in lower to mid forested areas, such as Hollyburn Mountain, Beaver Lake, Mundy Lake and Camosun Bog. Flowering starts mid-April; the fruit ripens as early as mid-June and continues to August at mid elevations.

SALMONBERRY
Rubus spectabilis • Rose family: *Rosaceae*

■ DESCRIPTION Salmonberry is one of B.C.'s tallest native berry bushes. Though it averages 2-3 m, the bush can grow up to 4 m high. The pink bell-shaped flowers, 3-4 cm across, bloom at the end of February and are a welcome sight. Flowering continues until June, when both the flowers and ripe fruit can be seen on the same bush. The soft logan-shaped berries range in colour from yellow to orange to red, with the occasional dark purple. The leaves are compound, with 3 leaflets, much like the leaves of a raspberry. Weak prickles may be seen on the lower portion of the branches; the tops are unarmed. The berry's common name comes from its resemblance to the shape and colour of salmon eggs.
■ HABITAT Common on the coast of B.C. in shaded damp forests.
■ NATIVE USE The high water content in the berries prevented them from being stored for any length of time. They were generally eaten shortly after harvesting.
■ LOCAL SITES Common in forested areas, such as Mundy Lake, Pacific Spirit Park and Minnekhada Park. Flowering starts at the end of February and the berries are harvested from the end of May to July.

KINNIKINNICK or BEARBERRY
Arctostaphylos uva-ursi • Heather family: *Ericaceae*

BERRIES

■ DESCRIPTION Kinnikinnick is a trailing, mat-forming evergreen that rarely grows above 25 cm in height. Its fragrant, pinkish flowers bloom in spring and are replaced by bright red berries 1 cm across by late summer. The small, oval leaves grow to 3 cm long, are leathery and alternate. Grouse and bears feed on the berries.

■ HABITAT Dry rocky outcrops and well-drained forest areas throughout B.C., from sea level to high elevations.

■ NATIVE USE *Kinnikinnick* is an eastern native word used to describe a tobacco mix. The leaves were dried and smoked, sometimes mixed with other plants.

■ LOCAL SITES Higher exposed areas from Shannon Falls to Mount Seymour and out to Harrison Lake at Bear Mountain. Good specimens can be seen at Lighthouse Park where it has also hybridized with manzanita (*A. columbiana*) to form manzanita cross (*A. x media*).

SALAL

Gaultheria shallon • Heather family: *Ericaceae*

■ **DESCRIPTION** Salal is a prostrate to mid-size bush that grows from 0.5 to 4 m in height. In spring the small pinkish flowers (1 cm long) hang like strings of tiny Chinese lanterns. The edible dark purple berries grow to 1 cm across and ripen by mid-August to September. Both the flowers and berries display themselves for several weeks. The dark green leaves are 7-10 cm long, tough and oval-shaped. Salal is often overlooked by berry pickers; the ripe berries taste excellent fresh and make fine preserves and wine.

■ **HABITAT** Dry to moist forested areas along the entire coast.

■ **NATIVE USE** Salal was an important food source for most native peoples. The berries were eaten fresh, mixed with other berries, or crushed and placed on skunk cabbage leaves to dry. The dried berry cakes were then rolled up and preserved for winter use.

■ **LOCAL SITES** Common understory bush; dense patches can be seen at Lighthouse Park, Alouette Lake and Harrison Lake. Flowering starts beginning of May, fruit starts to ripen beginning of August.

RED HUCKLEBERRY
Vaccinium parvifolium • Heather family: *Ericaceae*

BERRIES

■ DESCRIPTION One of the most graceful of all B.C.'s berry bushes, the red huckleberry grows on old stumps, where it can attain heights of 3-4 m. The combination of almost translucent red berries (1 cm across), lacy zigzag branch structure and pale green leaves (oval, 2-5 cm long) is unmistakable. The small greenish to pink flowers are inconspicuous.

■ HABITAT Coastal forested areas at lower elevations.

■ NATIVE USE The berries were eaten fresh, mixed with other berries and dried for winter use. Their resemblance to salmon eggs made them ideal for fish bait.

■ LOCAL SITES Common in forested areas, such as Lynn Headwaters Regional Park and Shannon Falls. Flowering starts mid-April and the berries ripen by the beginning of July.

OREGON GRAPE
Mahonia nervosa • Barberry family: *Berberidaceae*

■ **DESCRIPTION** Oregon grape is a smaller spreading understory shrub that is very noticeable when the bright yellow upright flowers are out. By midsummer the clusters of small green fruit (1 cm across) begin to turn an attractive grape blue. The leaves are evergreen, holly-like, waxy and compound, with usually 9-17 leaflets. The bark is rough, light grey outside and brilliant yellow inside. There is another species, tall Oregon grape (*M. aquifolium*), that grows in the same region. It is taller (2 m), has fewer leaflets (5-9) and prefers a more open and dry location. The genus name *Mahonia* commemorates American horticulturist Bernard McMahon, who died in 1816. The species name *aquifolium* means "holly-like."

■ **HABITAT** Dry coniferous forests in southern coastal B.C.

■ **NATIVE USE** When steeped, the shredded stems of both species yield a yellow dye that was used in basket-making. The tart berries were usually mixed with sweeter berries for eating.

■ **LOCAL SITES** Common understory plant in lower forested areas, such as Lighthouse Park and Pacific Spirit Park. Flowering starts at the end of March. The berries begin to turn blue by mid-July and persist through autumn.

LADY FERN
Athyrium filix-femina • Polypody family: *Polypodiaceae*

■DESCRIPTION Lady fern is a tall fragile fern to 2 m in height. The apple green fronds average up to 30 cm across and are widest below the centre, tapering at top and bottom. This diamond shape distinguishes the lady fern from the similar-looking spiny wood fern (*Dryopteris expansa*) whose fronds have an abrupt triangular form. The fronds die off in winter and emerge again in April. The horseshoe-shaped sori appear on the back of the fronds in spring.
■HABITAT Moist forests with nutrient-rich soils.
■NATIVE USE The young fronds (fiddleheads) were sometimes eaten in April.
■LOCAL SITES Common in damp forests at low to mid elevations, often in association with deer ferns and spiny wood ferns.

MAIDENHAIR FERN
Adiantum pedatum • Polypody family: *Polypodiaceae*

■DESCRIPTION Maidenhair fern is a delicate-looking fern with an almost tropical appearance. The fan-shaped fronds carry the dainty green leaflets (pinnules), which contrast well with the dark stems (stipes) that grow up to 60 cm in length. The reproducing sori under the pinnules are visible in late summer and fall. The genus name *Adiantum*, meaning "unwetted," refers to the way the fronds repel water.
■HABITAT Moist cliff faces at low to mid elevations.
■NATIVE USE The shiny black stipes were used in basket-making.
■LOCAL SITES Sporadic in Lighthouse Park. Lots can be seen carpeting the cliffs in Stanley Park and in higher areas along the Harrison River.

DEER FERN
Blechnum spicant • Polypody family: *Polypodiaceae*

■DESCRIPTION Deer fern can be distinguished from licorice fern (*Polypodium glycyrrhiza*) and sword fern (*Polystichum munitum*) by its two distinct types of frond, sterile and fertile. The sterile fronds grow up to 75 cm long, are tapered at both ends and usually lie flat. The fertile or spore-producing fronds are erect from the centre of the plant and can grow up to 75 cm in height. Deer ferns are good winter browse for deer.
■HABITAT Moist forested areas with plenty of rainfall.
■LOCAL SITES Not as common as sword fern but can be seen in most moist forested areas.

PARSLEY FERN or MOUNTAIN FERN
Cryptogramma crispa • Polypody family: *Polypodiaceae*

■ DESCRIPTION Parsley fern is a small evergreen to 30 cm in height. It has two sets of fronds, sterile and fertile (spore-producing). The sterile fronds are evergreen, deeply dissected, parsley-like and grow to 20 cm in height. The fertile fronds have less-congested foliage, their leaf margins are rolled and cover the sori, and they grow to 30 cm in height. The species name *crispa* refers to the crisped look of the fronds.

■ HABITAT Dry sites, typically rocky outcrops or slopes at low to high elevations.

■ LOCAL SITES North Shore mountains to Whistler and mountain ranges surrounding the Fraser Valley. The fertile fronds emerge in spring.

SPINY WOOD FERN OR SHIELD FERN
Dryopteris expansa • Polypody family: *Polypodiaceae*

■DESCRIPTION Spiny wood fern is an elegant plant to 1.5 m tall. The pale green fronds are triangular in shape, average up to 25 cm across and die off in winter. In spring the rounded sori are produced on the underside of the fronds. Spiny wood fern is similar in appearance and requirements to lady fern (*Athryium filix-femina*).
■HABITAT Common in moist forests at low to mid elevations.
■LOCAL SITES Common in cool, moist forested areas in the Lower Mainland. Young fronds can be seen emerging in early April.

WESTERN SWORD FERN
Polystichum munitum • Polypody family: *Polypodiaceae*

■DESCRIPTION Western sword fern is southern B.C.'s most common fern. It is evergreen and can grow to 1.5 m in height. The fronds are dark green with side leaves (pinnae) that are sharply pointed and toothed. On the underside of the fronds a double row of sori forms midsummer and turns orange by autumn. The fronds are in high demand in eastern Canada for floral decorations. The species name *munitum* means "armed," referring to the side leaves that resemble swords.

■HABITAT Dry to moist forest at lower elevations near the coast, where it can form pure groves.

■NATIVE USE The ferns were used to line steaming pits and baskets, and were placed on floors as sleeping mats.

■LOCAL SITES Common in all forested areas, especially in parks. New fronds emerge in April.

BRACKEN FERN
Pteridium aquilinum • Polypody family: *Polypodiaceae*

■ **DESCRIPTION** Bracken fern is B.C.'s tallest native fern, often reaching 3 m or more in height. It is also the most widespread fern in the world. The tall, arching fronds are dark green with a golden green stem (stipe), triangular in shape, and grow singly from rhizomes in spring.

■ **HABITAT** Has a diverse growing range, from dry to moist and open to forested regions.

■ **NATIVE USE** The rhizomes were peeled and eaten fresh or cooked, and the fiddleheads were boiled and eaten. However, it is not advisable to eat this fern as it has now been proven to be a health hazard.

■ **LOCAL SITES** Commonly seen from Harrison Hot Springs to Lynn Headwaters Regional Park to Shannon Falls.

LICORICE FERN
Polypodium glycyrrhiza • Polypody family: *Polypodiaceae*

■ **DESCRIPTION** Licorice fern is a smaller evergreen fern commonly seen on mossy slopes and on the trunks of bigleaf maple trees. The dark green fronds grow to 50 cm long and 5-7 cm wide and have a golden stem (stipe). The round spores are produced in a single row under the leaves. The rhizomes have a licorice taste, hence the fern's common name.

■ **HABITAT** Low-elevation forests, where it grows on trunks and branches of large trees, sometimes on shady outcrops.

■ **NATIVE USE** The roots were eaten fresh or cooked and were also used as a cold and throat medicine.

■ **LOCAL SITES** Commonly seen growing on the trunks of bigleaf maple trees. Lighthouse Park has sheets of them growing over its mossy rock faces.

COMMON HORSETAIL
Equisetum arvense • Horsetail family: *Equisetaceae*

ROGUES

■ DESCRIPTION Common horsetail is a herbaceous perennial to 75 cm in height. It has two types of stems, fertile and sterile, both hollow except at the nodes. The fertile stems are unbranched, to 30 cm in height, and lack chlorophyll; they bear spores in the terminal head. The green sterile stems grow to 75 cm in height and have leaves whorled at the joints. Horsetails are all that is left of a prehistoric family, some members of which grew to the size of trees.
■ HABITAT Low wet seepage areas, meadows, damp sandy soils and gravel roads from low to high elevations.
■ LOCAL SITES Damp ditches and moist neglected areas.

ROUND-LEAFED SUNDEW
Drosera rotundifolia • Sundew family: *Droseraceae*

■ DESCRIPTION There are about 100 species of sundew around the world, and all of them eat insects. B.C.'s native round-leafed sundew is a small perennial, 5-25 cm high, with inconspicuous white flowers. It is the leaves that make this plant a curiosity. They are equipped with fine red hairs, each tipped with a shiny globe of reddish secretion. Small insects are attracted to the secretion and get stuck in it; the leaf then slowly folds over and smothers the unsuspecting visitors. The plant's favourite foods are mosquitoes, gnats and midges.

■ HABITAT Peat bogs throughout the west coast of B.C.

■ NATIVE USE The whole plant is acrid, and the leaves were once used to remove corns, warts and bunions.

■ LOCAL SITES Spotty in Camosun Bog and Mundy Lake; carpets of sundew can be seen at Burns Bog.

CAT-TAIL
Typha latifolia • Cat-tail family: *Typhaceae*

■DESCRIPTION Cat-tails are semi-aquatic perennials that can grow to 2.5 m in height. The distinctive "tail," a brown spike, is 15-20 cm long and 3 cm wide and made up of male and female flowers. The lighter-coloured male flowers grow at the top and usually fall off, leaving a bare spike above the familiar brown female flowers. The sword-shaped leaves are alternate and spongy at the base.

■HABITAT Common in B.C. at low to mid elevations, at lakesides and riversides and in ponds, marshes and ditches.

■NATIVE USE The long leaves were used to weave mats and the fluffy seeds to stuff pillows and mattresses.

■LOCAL SITES Common in wet areas. Hectares of cat-tails can be seen on the banks of the Fraser River at the bottom of Boom Trail in Pacific Spirit Park. The spongy leaves are a favourite food of the beavers in the Jericho Beach Park pond.

RUNNING CLUBMOSS
Lycopodium clavatum • Clubmoss family: *Lycopodiaceae*

■ DESCRIPTION Running clubmoss is a curious creeping evergreen that looks like it is made of bright green pipe cleaners. Like all clubmosses, it has no flowers and reproduces by spores. These are held in terminal cones on vertical stalks to 25 cm in height. The evergreen leaves are lance-shaped and arranged spirally around the stem. Running clubmoss grows horizontally across the ground, with irregular rooting. The spores are used medicinally and in industry.

■ HABITAT Dry to moist coniferous forests at low to high elevations.

■ LOCAL SITES Common across the North and West Vancouver mountains. Can be seen at the top of the Grouse Grind and at Cypress Bowl.

STINGING NETTLE
Urtica dioica • Nettle family: *Urticaceae*

■ DESCRIPTION Stinging nettle is a herbaceous perennial to over 2 m in height. Its tiny flowers are greenish and produced in hanging clusters to 5 cm long. The leaves are heart-shaped at the base, tapered to the top, coarsely toothed, to 10 cm long. The stalks, stems and leaves all have stinging hairs that contain formic acid; many people have the misfortune of encountering this plant the hard way. The genus name *Urtica* is from the Latin *uro* ("to burn").

■ HABITAT Thrives in moist, nutrient-rich, somewhat shady disturbed sites, where it can form great masses. Stinging nettles are usually an indicator of nitrogen-rich soil.

■ NATIVE USE The young leaves were boiled as a spinach substitute.

■ LOCAL SITES Scattered throughout the Lower Mainland, with huge patches at Caulfeild Cove and the forest edges of Pacific Spirit Park. Flowering starts at the beginning of May.

INDIAN PIPE
Monotropa uniflora • Indian Pipe family: *Monotropaceae*

■ **DESCRIPTION** Indian Pipe is a fleshy ghost-like herbaceous perennial to 25 cm in height. It has a single white flower (hence the species name, *uniflora*), to 3 cm long; this starts off nodding but stands erect when mature. The leaves are white, scale-like and clasp the stem. The entire plant turns black at the end of the growing season. The plant is usually seen growing in clumps of up to 25 stems.

■ **HABITAT** Coniferous forests with nutrient-rich soil and deep shade at low elevations.

■ **LOCAL SITES** Not commonly seen, though it can be found at Alouette Lake and Mundy Lake. Flowers throughout July, with the black stems and seed heads persisting into autumn.

DEER CABBAGE
Fauria crista-galli • Buck bean family: *Menyanthaceae*

ROGUES

■ **DESCRIPTION** Deer cabbage is a low-growing lush perennial to 50 cm in height. Its white flowers are split into 5 wavy lobes (petals), to 2 cm across, and sit in open clusters on stems 20-50 cm long. The leaves are basal, 7-12 cm across, rounded to heart-shaped, with bumpy edges (crenate). The species name *crista-galli* means "cockscomb," a reference to the wavy lobes.

■ **HABITAT** Moist to wet forests, bogs and seepage areas at low to high elevations.

■ **LOCAL SITES** Common at mid levels on local mountains. Large patches between First Lake on Hollyburn Mountain and the Cypress Bowl parking lot. Flowers end of June through July.

SCOURING RUSH
Equisetum hyemale • Horsetail family: *Equisetaceae*

■DESCRIPTION Scouring rush is a herbaceous perennial to 1.7 m in height. Its dark green stems are all alike, with black rings separating the hollow sections. They are branchless and rough or scratchy to touch. Like ferns, horsetails do not produce flowers or fruit but reproduce from spores. These are borne in hard, pointed terminal cones.

■HABITAT Usually close to fresh water, by streams and rivers or at the base of moist slopes with loose rich soil.

■NATIVE USE The abrasive stems were used as sandpaper and the dark roots to make baskets.

■LOCAL SITES Extensive patches at the base of Boom Trail in Pacific Spirit Park and scattered patches farther up the Fraser River. Not as widespread as common horsetail.

NOOTKA ROSE
Rosa nutkana • Rose family: *Rosaceae*

■DESCRIPTION The largest of B.C.'s native roses, the Nootka rose grows to 3 m in height. The showy pink flowers are 5-petalled, fragrant, 5 cm across and usually solitary. The compound leaves have 5-7 toothed leaflets and are armed with a pair of prickles underneath. The reddish hips are round and plump, 1-2 cm across, and contrast well with the dark green foliage.

■HABITAT Open low-elevation forests throughout B.C.

■NATIVE USE Rosehips were strung together to make necklaces and the flowers were pressed to make perfume. Rosehips were only eaten in times of famine.

■LOCAL SITES Rocky outcrops in lower North and West Vancouver, from Lighthouse Park to Shannon Falls and at Noons Creek in Port Moody. Burns Bog has an excellent flower display in early June. Flowers begin mid-May and the hips start to develop colour by the beginning of August.

BALDHIP or WOODLAND ROSE
Rosa gymnocarpa • Rose family: *Rosaceae*

■DESCRIPTION The baldhip rose is B.C.'s smallest native rose. It is often prostrate, to 1.4 m in height. The tiny pink flowers are 5-petalled, delicately fragrant, 1-2 cm across and usually solitary. The compound leaves are smaller than the Nootka rose and have 5-9 toothed leaflets. The spindly stems are mostly armed with weak prickles. A good identifier is this rose's unusual habit of losing its sepals, leaving the hip bald — hence the species name *gymnocarpa*, which means "naked fruit." Rosehips have a higher concentration of vitamin C than oranges and make an excellent jelly or marmalade.
■HABITAT Dry open forests at lower elevations, from southern B.C. to the redwood forests of California.
■LOCAL SITES Not as common as the Nootka rose; can be seen sporadically on low rock outcrops from Shannon Falls to West Vancouver, and out to the Fraser Valley. Flowering starts at the end of May, with a good fruit crop by mid-July.

HARDHACK or STEEPLEBUSH
Spiraea douglasii • Rose family: *Rosaceae*

■DESCRIPTION Hardhack is an upright deciduous bush to 2 m in height. Its tiny pink flowers group together to form fuzzy pyramidal clusters up to 15 cm tall. The resulting brown fruiting clusters persist on the bush through winter. The alternate leaves are elliptic to oval, toothed above the middle, 5-10 cm long, dark green above and a felty paler green below.

■HABITAT Prefers moist conditions, can be seen growing in ditches and bogs and at lakesides from low levels to sub-alpine meadows.

■NATIVE USE The tough wiry branches were used to make halibut hooks, scrapers and hooks for drying and smoking salmon.

■LOCAL SITES Common in moist areas throughout the Lower Mainland. Impenetrable thickets can be seen at Burns Bog. Flowering begins mid-June and fades by the beginning of August.

OCEANSPRAY
Holodiscus discolor • Rose family: *Rosaceae*

■DESCRIPTION Oceanspray is an upright, deciduous shrub to 5 m in height. Its small, creamy-white flowers are densely packed to form reversed pyramidal clusters to 20 cm long. The fruiting clusters turn an unattractive brown and persist through winter. The leaves are wedge-shaped, flat green above, pale green and hairy below, to 5 cm long. The species name *discolor* refers to the two-coloured leaf.
■HABITAT Dry open forests at low to mid elevations; often found on rocky outcrops.
■NATIVE USE The straight new growth was a favourite for making arrows, hence its other name, arrow-wood. The wood is extremely hard and was used to make harpoon shafts, teepee pins, digging tools and drum hoops.
■LOCAL SITES Common in open dry areas, especially near the ocean. Abundant above Wreck, Dunbar and Trafalgar beaches and in the Harrison Lake area (Bear Mountain trails). In full flower by the end of June.

INDIAN PLUM
Oemleria cerasiformis • Rose family: *Rosaceae*

■ **DESCRIPTION** Indian plum is an upright deciduous shrub or small tree to 5 m in height. Its flowers, which usually emerge before the leaves, are white, 1 cm across, and hang in clusters 6-10 cm long. The small, plum-like fruit grow to 1 cm across; they start off yellowish red and finish a bluish black. They are edible, but a large seed and bitter taste make them better left for the birds. The leaves are broadly lance-shaped, light green, 7-12 cm long, and appear in upright clusters. The species name *cerasiformis* means "cherry-shaped," a reference to the fruit.

■ **HABITAT** Restricted to low elevations on the southern coast and Gulf Islands; prefers moist, open broad-leafed forests.

■ **NATIVE USE** Small amounts were eaten fresh or dried for winter use.

■ **LOCAL SITES** In early March, Indian plum flowers light up forest edges along Pacific Spirit Park, Jericho Beach Park and Hwy 1 from Burnaby to Hope. Yellow fruit on the female bushes are seen by the end of May, with ripe fruit by the end of June.

GOAT'S BEARD
Aruncus dioicus • Rose family: *Rosaceae*

■DESCRIPTION Goat's beard is a deciduous shrub to 3 m in height. The plants are dioecious — male and female flowers appear on separate plants. The tiny white flowers are compacted into hanging panicles up to 60 cm long. The leaves are compound 3 times (thrice pinnate); leaflets are bright green with a toothed edge, tapering to a point. With a little imagination, the hanging flowers can look like a goat's beard.
■HABITAT Moist open woodlands, creeksides and wet rocky slopes at lower elevations.
■LOCAL SITES Common throughout the Lower Mainland. At the end of May masses of flowers can be seen on the sides of the Sea to Sky Highway (Hwy 99).

RED-BERRIED ELDER or RED ELDERBERRY
Sambucus racemosa • Honeysuckle family: *Caprifoliaceae*

■DESCRIPTION Red-berried elder is a bushy shrub to 6 m in height. Its small flowers are creamy-white and grow in pyramidal clusters 10-20 cm long. The berries that replace them take up to 3 months to turn bright red; they are considered poisonous to people when eaten raw but are a favourite food for birds. The leaves are compound, 5-15 cm long, with 5-9 opposite leaflets.
CAUTION: the berries are considered poisonous.

■HABITAT Moist coastal forest edges and roadsides. The blue-berried elder (*S. caerulea*) is found more in the Interior and the Gulf Islands.

■NATIVE USE The pithy branches were hollowed out and used as blowguns.

■LOCAL SITES Common in lower forested areas; thick patches in Stanley Park, Lighthouse Park, Debouville Slough, Alouette Lake and Harrison Lake. Flowering starts mid-April, berries turn bright red late June-early July at lower elevations.

SNOWBERRY
Symphoricarpos albus • Honeysuckle family: *Caprifoliaceae*

■DESCRIPTION Snowberry is an erect deciduous shrub to 2 m in height. Its small white to pink flowers turn into an abundance of very noticeable white berries 1-2 cm across. Older plants produce a smaller oval leaf to 2 cm long, while younger, more vigorous plants have wavy leaves to 5 cm long. The leaves have a sweet fragrance when wet. Snowberries are best appreciated in winter, when the bright white berries stand out against the surrounding grey. The genus name *Symphoricarpos* refers to the clustering of the berries.

CAUTION: the berries are considered poisonous.

■HABITAT Open forested areas at low to mid elevations.

■NATIVE USE Thin branches were hollowed out to make pipe stems and larger branches were bound together to make brooms.

■LOCAL SITES Common in lower forested areas; good concentrations on SW Marine Drive through Pacific Spirit Park, on the edges of Harrison Lake and in Central Park in Burnaby. Flowers mid-June, with the berries noticeable by September.

BLACK TWINBERRY
Lonicera involucrata • Honeysuckle family: *Caprifoliaceae*

■ DESCRIPTION Black twinberry is a deciduous shrub 1-3 m in height. Its yellow flowers are tubular and borne in pairs, to 2 cm long. The inedible berries are shiny black, cupped in a moon bract, to 1 cm across. The leaves are broadly lance-shaped, tapering to a point, opposite, 5-15 cm long.

■ HABITAT Moist to wet open forests at low to high elevations.

■ NATIVE USE The berries were mashed and the purple juice used to dye roots for basketry. The Haida rubbed the berries into their scalps to prevent their hair from turning grey. It was said that eating the berries drove a person crazy.

■ LOCAL SITES Common in moist areas. Good quantities can be seen at Minnekhada Park, Noons Creek in Port Moody, Deer Lake, Burns Bog, Stanley Park above the seawall and from Caulfeild Cove to Shannon Falls. Flowering starts mid-May, with the berries appearing mid-June.

FALSE AZALEA or FOOL'S HUCKLEBERRY
Menziesia ferruginea • Heather family: *Ericaceae*

■DESCRIPTION False azalea is an upright, deciduous shrub to 3 m in height. Its flowers, which resemble a huckleberry's, are a dull copper colour, bell-shaped, to 8 mm long, with long stems. The small fruit (5 mm long) is a dry, four-valved capsule that is not edible. The leaves are elliptic, bluish green on top, whitish green below, 3-6 cm long. They appear to grow in whorls. The genus name *Menziesia* is after Archibald Menzies, a naval surgeon and botanist who sailed with Captain George Vancouver and collected plants on the West Coast.

■HABITAT Moist forested sites, especially in wetter areas, at low to high elevations.

■LOCAL SITES Common on mountain slopes from the North Shore to Whistler. Large specimens can be seen just past the boardwalk at Noons Creek in Port Moody and on the trail up to Stawamus Chief above Squamish.

WESTERN BOG LAUREL
Kalmia microphylla ssp. *occidentalis* • Heather family: *Ericaceae*

■ **DESCRIPTION** B.C.'s native laurel is a small lanky evergreen no more than 60 cm high. Its beautiful pink flowers (2.5 cm across) have a built-in pollen dispenser. A close look reveals that some of the stamens are bent over; these spring up when the flower is disturbed, dusting the intruder with pollen. The leaves are opposite, lance-shaped, 2-4 cm long, shiny dark green above, felty white below, with edges strongly rolled over. The plant is poisonous and should not be confused with Labrador tea (*Ledum groenlandicum*; see opposite), which it resembles from above. CAUTION: the plant is poisonous.

■ **HABITAT** Peat bogs and lakeshores at low to high elevations throughout B.C.

■ **NATIVE USE** The leaves were boiled and used in small doses for medicinal purposes.

■ **LOCAL SITES** Camosun Bog, Burns Bog and Mundy Lake. Flowers early May.

LABRADOR TEA
Ledum groenlandicum • Heather family: *Ericaceae*

■DESCRIPTION Most of the year, Labrador tea is a gangly small shrub to 1.4 m in height. In spring the masses of small white flowers turn it into the Cinderella of the bog. The evergreen leaves are lance-shaped, alternate, 4-6 cm long, with the edges rolled over. The leaves can be distinguished from those of the poisonous bog laurel (*Kalmia microphylla*; see opposite) by their flat green colour on top and rusty-coloured hairs beneath. To be safe, only pick the leaves when the shrub is in flower.

■HABITAT Peat bogs, lakesides and permanent wet meadows, low to alpine elevations.

■NATIVE USE The leaves have long been used by native groups across Canada as infusions. Early explorers and settlers quickly picked up on this caffeine substitute. Caution must be taken – not all people can drink it.

■LOCAL SITES Deer Lake, Burns Bog, Camosun Bog and Mundy Lake. Flowering starts mid-May, best viewing is end of May, early June.

SCOTCH BROOM
Cytisus scoparius • Pea family: *Leguminosae*

■ **DESCRIPTION** Scotch broom is a mid-size, unarmed, deciduous shrub to 3 m in height. The bright yellow, pea-like flowers are 2 cm long and project at all angles from the stems. When the flowering period is over, the bushes are draped with thousands of brown-black pods up to 5 cm long. These pods split open when dry, forcing the seeds several metres away from the parent plant. The leaves are short and narrow and pressed close to the stems. A native of Scotland, broom was introduced near Victoria in the mid-1800s and has since become an invasive weed on southern Vancouver Island and the adjoining coast. As its common name suggests, its branches were tied together and used as a broom.
■ **HABITAT** Dry sites, rocky outcrops, roadsides and parkland.
■ **LOCAL SITES** Common; dense patches can be seen on exposed rocky areas in Lighthouse Park and beside the Sea to Sky Highway (Hwy 99) and Hwy 1. Flowering starts early April.

RED-FLOWERING CURRANT
Ribes sanguineum • Currant family: *Grossulariaceae*

■DESCRIPTION Red-flowering currant is a deciduous upright bush from 1.5-3 m in height. Its flowers range in colour from pale pink to bright crimson and hang in 8-12 cm panicles. The bluish black berries (1 cm across) are inviting to eat but are usually dry and insipid. The leaves are 5-10 cm across, maple-shaped, with 3-5 lobes. Currants and gooseberries are in the same genus (*Ribes*); a distinguishing feature is that currants have no prickles, while gooseberries do.

■HABITAT Dry open coastal forests at low to mid elevations.

■LOCAL SITES Cypress Park, Hollyburn Mountain, Lighthouse Park and Stanley Park. Flowering starts mid-March to late April, often in association with skunk cabbage.

DEVIL'S CLUB
Oplopanax horridus • Ginseng family: *Araliaceae*

■ **DESCRIPTION** Devil's club is a deciduous shrub from 1-3 m in height. Its small white flowers are densely packed into pyramidal clusters approximately 15 cm long. The flowers bloom in May and are replaced by showy scarlet berries in August; these are not considered edible. The large leaves are maple-like, alternate, to 30 cm across, with spines in the larger veins on both sides. The stems are sprawling, awkward-looking and very well-armed with spines to 1 cm long. The species name *horridus* comes to mind when you accidentally encounter this shrub.

■ **HABITAT** Moist forested areas with rich soil at low to mid elevations.

■ **NATIVE USE** Next to hellebore, devil's club was coastal natives' most valued medicinal plant. Infusions and poultices were used to relieve arthritis, fevers, colds and infections.

■ **LOCAL SITES** Forested ravines by the Harrison River, Alouette Lake and in Port Coquitlam, large concentrations from Brandywine Falls to Garibaldi and Whistler.

RED-OSIER DOGWOOD
Cornus stolonifera • Dogwood family: *Cornaceae*

■ DESCRIPTION Red-osier dogwood is a mid-size deciduous shrub to 5 m in height. Its small white flowers (4 mm across) are grouped together to form dense round clusters approximately 10 cm across. By August they have been replaced by bunches of dull white inedible berries to 8 mm across. The leaves are typically dogwood: opposite, to 10 cm long, with parallel veins. Younger branches are pliable and have an attractive red colour.

■ HABITAT Moist to wet areas, usually forested, at low to mid elevations.

■ NATIVE USE The small branches were used for weaving, barbecue racks, fuel for smoking salmon and latticework for fishing weirs.

■ LOCAL SITES Common, but often hiding among other bushes. Good concentrations at Debouville Slough, Burns Bog, Noons Creek in Port Moody, Pacific Spirit Park and along the Fraser River all the way from the Musqueam Reserve to Harrison Lake. Flowering starts late May, berries ripen end of July to August.

FALSEBOX or MOUNTAIN LOVER
Pachistima myrsinites • Staff tree family: *Celastraceae*

■DESCRIPTION Falsebox is a small evergreen shrub to 75 cm in height. Its tiny maroon flowers go unnoticed by all but a curious few. The evergreen leaves are elliptic, leathery, tooth-edged, to 3 cm long. Falsebox is an attractive bush more noticed for its foliage than its flowers. The species name *myrsinites* is Greek for myrrh, in reference to the fragrant flowers.

■HABITAT Forested mountain slopes at low to mid elevations. Rare on the coast, but abundant in the B.C. interior.

■LOCAL SITES Common from North and West Vancouver mountains to Whistler and in dry mountain areas surrounding the Fraser Valley. Flowering starts at the end of March.

SWEET GALE
Myrica gale • Wax myrtle family: *Myricaceae*

■DESCRIPTION Sweet gale is an aromatic deciduous shrub to 1-3 m in height. Its flowers are displayed in yellowish male and female catkins, which appear in spring before the leaves. The fruit are tiny brown cone-like husks that persist through the winter. The thin leaves are flat green above and whitish below, coarsely toothed above the middle, to 5 cm long. The genus name *Myrica* means "perfume," a reference to the sweet-scented leaves.

■HABITAT Prefers to have its feet wet in shallow fresh water: ponds, lakes and swamps at low to mid elevations.

■LOCAL SITES Deer Lake and Burns Bog; when out of flower it blends in with hardhack (*Spiraea douglasii*: see page 99).

PINK MOUNTAIN HEATHER
Phyllodoce empetriformis • Heather family: *Ericaceae*

■**DESCRIPTION** Pink mountain heather is a mat-forming evergreen shrub 10-50 cm in height. Its rose pink flowers are bell-shaped to 1 cm long and are held out on long, slender stalks. With their stiff needle-like leaves to 1 cm long, the branches resemble miniature conifers. A hike into alpine meadows is well worth the effort to see this beautiful fragrant plant.

■**HABITAT** Rocky sites at sub-alpine to alpine elevations.

■**LOCAL SITES** Top of the Grouse Grind, Hollyburn Mountain and exposed areas at upper elevations to Whistler. Flowers mid-June through July locally, beginning of August in the alpine meadows at Whistler/Blackcomb.

NINEBARK
Physocarpus capitatus • Rose family: *Rosaceae*

■**DESCRIPTION** Ninebark is an upright deciduous shrub to 4 m in height. Its tiny white flowers are grouped into rounded clusters to 7 cm across. The fruit are reddish brown inflated seed capsules. The maple-shaped leaves are 3-5 lobed, shiny green above, paler below, to 7 cm long. It is debatable whether the shaggy bark has nine layers. The species name *capitatus* refers to the rounded heads of the flowers.
■**HABITAT** Usually seen on moist sites, in open forests and at streams and lakesides, but also on dry rocky areas at lower elevations.
■**NATIVE USE** Ninebark was used medicinally.
■**LOCAL SITES** Lighthouse Park, Cypress Park, Minnekhada Park and moist sites out the Fraser Valley to Harrison Lake.

VINE MAPLE
Acer circinatum • Maple family: *Aceraceae*

■DESCRIPTION Vine maple is a shrub or small tree 3-9 m in height. It is often multi-stemmed, with older trees occasionally becoming prostrate. The leaves resemble Japanese maples, with 7-9 serrated lobes, opposite, 7-13 cm across. The winged seeds are 2-4 cm long, starting green and becoming reddish brown by autumn. The bark is smooth and pale green on young trees, dull brown on older ones. The autumn leaves create an incredible colour show.
■HABITAT Moist shaded woods at lower elevations.
■NATIVE USE The wood was used to make bows, arrows, spoons, handles and snowshoe frames.
■LOCAL SITES Common understory tree, very noticeable October to November. Both sides of the road from Prospect Point to Third Beach in Stanley Park are lined with 9-m vine maples; the drive is well worth taking in autumn.

BIGLEAF MAPLE
Acer macrophyllum • Maple family: *Aceraceae*

■DESCRIPTION The bigleaf maple is the largest native maple on the West Coast, often exceeding heights of 30 m. Its huge leaves, which are dark green, 5-lobed, 20-30 cm across, are excellent identifiers. In early spring it produces beautiful clusters of scented yellow-green flowers, 7-10 cm long. The mature winged seeds (samaras), 5 cm long, act as whirlygigs when they fall; they are bountiful and an important food source for birds, squirrels, mice and chipmunks. The brown fissured bark is host to an incredible number of epiphytes, most commonly mosses and licorice ferns (see page 87).

■HABITAT Dominant in lower forested areas. Its shallow root system prefers moist soils, mild winters and cool summers.

■NATIVE USE The plentiful wood was important in native culture, as a fuel and for carvings, paddles, combs, fish lures, dishes and handles. The large leaves were used to line berry baskets and steam pits.

■LOCAL SITES Common in lower forested areas. Large specimens can be seen at Harrison Lake, Alouette Lake and in Pacific Spirit Park. B.C.'s tallest recorded bigleaf maple, 31.7 m, is in Stanley Park. Flowers mid-April.

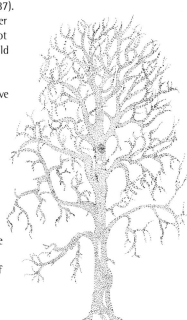

RED ALDER
Alnus rubra • Birch family: *Betulaceae*

■DESCRIPTION The largest native alder in North America, the red alder grows quickly and can reach 25 m in height. Its leaves are oval-shaped, grass green, 7-15 cm long, with a coarsely serrated edge. Hanging male catkins, 7-15 cm long, decorate the bare trees in early spring. The fruit (cones) are 1.5 to 2.5 cm long; they start off green, then turn brown and persist through winter. The bark is thin and grey on younger trees, scaly when older. Red alder leaves give a poor colour display in autumn, mainly green or brown.

■HABITAT Moist wooded areas, disturbed sites and stream banks at low to mid elevations.

■NATIVE USE The soft straight-grained wood is easily worked and was used for making masks, bowls, rattles, paddles and spoons. The red bark was used to dye fishnets, buckskins and basket material.

■LOCAL SITES Very common in forested areas; the northern portion of Pacific Spirit Park, for instance, is dominated by hectares of 15- to 20-m red alders. B.C.'s tallest recorded red alder, 41 m, is at Third Beach in Stanley Park.

< *The Third Beach red alder (top)*

SITKA ALDER
Alnus sinuata • Birch family: *Betulaceae*

■DESCRIPTION The Sitka alder is a deciduous shrub or small tree 3-7 m high. Its coarse leaves are double-serrated, grass green and 7-10 cm long. In early spring it becomes covered in pollen-producing catkins 10-15 cm long and female cones 2 cm long.

■HABITAT As with most alders, it prefers moist conditions, from the coast of the Arctic Circle to the high mountains of California. Often grows on avalanche sites.

■NATIVE USE The soft straight-grained wood is easily worked and was used for making masks, bowls, rattles, paddles and spoons. The red bark was used to dye fishnets, buckskins and basket material.

■LOCAL SITES Commonly seen at upper elevations from Grouse Mountain to Shannon Falls. Also scattered throughout Lighthouse Park.

PACIFIC CRAB APPLE
Malus fusca • Rose family: *Rosaceae*

■**DESCRIPTION** Pacific crab apple is a deciduous shrub or small tree 2-10 m high. Its leaves (5-10 cm long) are similar to those of orchard apple trees, except that they often have bottom lobes. The flowers are typical apple blossoms, white to pink, fragrant and in clusters of 5-12. The fruit that follows is 1-2 cm across, green at first, turning yellowy reddish. On older trees the bark is scaly and deeply fissured. The Pacific crab apple is B.C.'s only native apple.

■**HABITAT** High beaches, moist open forests, swamps and stream banks at lower elevations.

■**NATIVE USE** The small apples were an important food source, and the hard wood was used to make digging sticks, bows, handles and halibut hooks.

■**LOCAL SITES** Lighthouse Park, cliffs above the Stanley Park seawall and at Mundy Lake, Noons Creek in Port Moody, Minnekhada Park and Harrison River. Flowering starts mid-May.

ARBUTUS or PACIFIC MADRONE
Arbutus menziesii • Heather family: *Ericaceae*

■DESCRIPTION The arbutus is often seen as a contorted shrub or small tree, but in ideal conditions it can attain heights of 30 m. Its rhododendron-like leaves are leathery, glossy dark green above, silvery white below, to 15 cm long. The panicles of white flowers produced in spring are followed by clusters of orange red berries 1 cm across. The reddy brown bark peels away every year to reveal a greenish underbark; the beautiful colours of the thin peeling outer bark are the key identifiers. The arbutus is Canada's only native broadleaf evergreen tree.

■HABITAT Dry coastal forests on the Gulf Islands, southeastern Vancouver Island and the adjacent coast. Southwestern B.C. is the northern extent of arbutus.

■NATIVE USE The leaves and bark were steeped as tea and used to cure colds and stomach aches. The wood cracks very easily and was rarely used to make tools or carvings.

■LOCAL SITES Lower elevations in West Vancouver and up Howe Sound past Lions Bay; the best places to see arbutuses are Whytecliff Park and Lighthouse Park. Flowering starts mid-April. The tallest known arbutus in B.C. can be found on Thetis Island and is 31.7 m high.

PAPER BIRCH or CANOE BIRCH
Betula papyrifera • Birch family: *Betulaceae*

■ DESCRIPTION Paper birch is a medium-size tree reaching heights of 20 m. Its serrated leaves are 8-12 cm long, rounded at the bottom and sharply pointed at the apex. Male and female catkins can be seen in early spring just before the leaves appear. The white peeling bark is a good identifier on younger trees. There is a red-bark variety that can be confused with the native bitter cherry (*Prunus emarginata*). The species name *papyrifera* means "to bear paper."

■ HABITAT Rare in low-elevation coastal forests, common in interior forests; prefers moist soil and will tolerate wet sites.

■ NATIVE USE The bark was used to make canoes, cradles, food containers, writing paper and coverings for teepees. The straight-grained wood was used for arrows, spears, snowshoes, sleds and masks.

■ LOCAL SITES Scattered throughout lower forested areas. Larger trees can be seen at Mundy Lake, Deer Lake, Pacific Spirit Park, Locarno Beach and Harrison Lake. The tallest known paper birch in B.C., 25.9 m, is near Pemberton.

PACIFIC DOGWOOD
Cornus nuttallii • Dogwood family: *Cornaceae*

■**DESCRIPTION** Pacific dogwood ranges from being a multi-trunked shrub to a medium-size tree as tall as 19 m. Its leaves are elliptical, deep green above, lighter green below, to 10 cm long. The flowers are not quite as they seem: the 4-7 showy white petals are actually bracts that surround small, greenish white flowers. Clusters of small red berries 1 cm across appear by late summer. The bark is dark brown and smooth on young trees, scaly and ridged on older ones. The flower is the floral emblem of B.C., and the tree is protected by law. The painter and ornithologist John James Audubon named this tree after his friend Thomas Nuttall, the first person to classify it as a new species.

■**HABITAT** Coastal forests at low elevations.

■**NATIVE USE** The hard wood was used to make bows, arrows, harpoon shafts and, more recently, knitting needles.

■**LOCAL SITES** Sporadic in forested areas in the Fraser Valley and at lower elevations from West Vancouver to Squamish; Lighthouse Park has some old specimens. Probably seen more in cultivation than in the wild. Flowering starts in mid-April and there is often a second, smaller show in September.

COTTONWOOD
Populus balsamifera ssp. *trichocarpa* •
Willow family: *Salicaceae*

■DESCRIPTION Cottonwood is the tallest deciduous tree in the Pacific Northwest. It is also one of the fastest-growing, attaining heights of 45 m and trunk diameters of 2-3 m. Its leaves are heart-shaped, alternating, dark shiny green above, pale green below, 7-15 cm long. The tiny seeds on female trees hang on 7- to 13-cm catkins and are covered in white fluffy hairs known as "cotton." The deeply furrowed bark and large sticky buds are good identifiers in winter. In early summer bits of cotton can be seen filling the skies, transporting the seeds many kilometres away from the parent trees. The wood is used commercially to make tissue paper.

■HABITAT Low moist to wet areas across B.C. Requires sunshine and will not tolerate heavy shade.

■NATIVE USE The sticky gum from the buds was boiled and used to stick feathers on arrow shafts and to waterproof baskets and birchbark canoes.

■LOCAL SITES Common in moist areas. It is a dominant deciduous tree around Chilliwack, Agassiz and Harrison Lake.

BITTER CHERRY
Prunus emarginata • Rose family: *Rosaceae*

■DESCRIPTION Bitter cherry is a small to medium-size tree, 5-15 m in height. Its bark is very distinctive: reddy brown with orange slits (lenticels), it is thin, smooth and peels horizontally. The white flowers (1 cm across) put on a wonderful show in April. The immature green fruit, 1 cm across, appear in early June and are bright red by midsummer. The leaves are quite different from those of the familiar Japanese cherries; they are alternate, 3-8 cm long, very finely toothed and usually rounded at the tip. The cherries are extremely bitter and considered inedible to humans, but are an important food source for birds. The fruit pits do not break down when digested, so birds carry them many kilometres from the parent tree.

■HABITAT Scattered in disturbed forests at low to mid elevations.

■NATIVE USE The shiny red bark was used to make baskets, mats and bags. The hard wood makes excellent fuel.

■LOCAL SITES Common in lower forested areas, such as the forest edges of Hwy 1, Pacific Spirit Park and Stanley Park; very noticeable when flowering in mid-April.

CASCARA
Rhamnus purshiana • Buckthorn family: *Rhamnaceae*

■DESCRIPTION Cascara can be a multi-trunked shrub or a small tree to 9 m in height. Its leaves are oblong with prominent veins, glossy, grass green, 7-13 cm long. The flowers are small, greenish yellow and rather insignificant. The berries look like small cherries, 1-8 mm across, red at first, turning blue black in late summer. The smooth bark is silver grey and on older trees it resembles an elephant's hide. The bark was collected commercially for years and used as the key ingredient in laxatives.

■HABITAT Prefers moist, nutrient-rich sites in the shade of larger trees at low elevations.

■NATIVE USE The bark was boiled and the infusion used as a laxative.

■LOCAL SITES Common in lower forested areas; large trees can be seen in Pacific Spirit Park and Lighthouse Park.

WILLOWS
Salix sp. • Willow family: *Salicaceae*

■DESCRIPTION Native willows are easy to identify as a genus but hard to distinguish as species. This is due to the variable leaf shapes within the same species, male and female flowers on separate plants, flowering before leaves appear and hybridization between species. The two most common willows are:

Scouler's willow (*Salix scouleriana*) – a shrub or tree 5-12 m in height, leaves 5-8 cm long, felty, narrow at the base and rounded at the tip. The flowers appear before the leaves, males to 4 cm long, females to 6 cm.
■HABITAT Scattered in disturbed spots in young forests at low to mid elevations.

Pacific willow (*Salix lucida* ssp. *lasiandra*) – a shrub or tree 6-12 m in height, leaves 10-15 cm long, lance-shaped, with finely toothed edges. The flower appears with the leaves, males to 7 cm long, females to 12 cm.
■HABITAT Commonly seen in shallow fresh water, ditches, creeks and at lakesides. Low to mid elevations.

Scouler's Willow

Pacific Willow

Pacific Willow >

SITKA MOUNTAIN ASH
Sorbus sitchensis • Rose family: *Rosaceae*

■DESCRIPTION Sitka mountain ash is a small, multi-stemmed bush or thicket from 1.5 to 4.5 m in height. Its compound bluish green leaves have 7-13 leaflets, 11 being the norm. The tiny white flowers are in terminal clusters, 5-10 cm across. In August and September the bushes and trees display a wonderful show of bright red-orange berries. The bark is thin and shiny grey. The native mountain ash should not be confused with the larger European mountain ash (*S. Aucuparia*), an introduced species that has naturalized well as its berries are a favourite with birds. The European ash, or rowan, is rich in history. In Britain it was planted near homes to protect owners from witches and in cemeteries to keep the dead in their graves. Christ is believed to have been crucified on a cross made from mountain ash, cedar, holly pine or cypress.

■HABITAT Native ash stays primarily where its name suggests – in the mountains. The European ash can be found at lower elevations, especially near townships.

■LOCAL SITES Top of the Grouse Grind, on Hollyburn Mountain, in the Harrison Lake area and elsewhere at elevations similar to Whistler Mountain.

BLACK HAWTHORN
Crataegus douglasii • Rose family: *Rosaceae*

■ DESCRIPTION Black hawthorn is an armed, scraggly shrub or bushy tree to 9 m in height. The leaves are roughly oval, coarsely toothed above the middle to 6 cm long. The clusters of white flowers are showy but bland in smell. The edible fruit is purple black to 1 cm long and hangs in bunches. Older bark is grey, patchy and very rough.

■ HABITAT Prefers moist soil beside streams, in open forests or near the ocean.

■ NATIVE USE The 3-cm thorns were used as tines on herring rakes.

■ LOCAL SITES Scattered along the banks of the Fraser River and in Minnekhada Park. Flowers in April to May.

PACIFIC SILVER FIR
Abies amabilis • Pine family: *Pinaceae*

■ DESCRIPTION Pacific silver fir is a straight-trunked, symmetrical conifer to 60 m in height. The bark on young trees is smooth grey, with prominent vertical resin blisters. As the tree ages the bark becomes scaly, rougher and often lighter in colour. The cones are dark purple, barrel-shaped to 12 cm long, and sit erect on the upper portion of the tree. The needles are a lustrous dark green on the upper surface, silvery white below, with a notched tip. Pacific silver fir is one of Canada's stateliest conifers. The species name *amabilis* means "lovely fir."

■ HABITAT Moist forests at mid to high elevations.

■ NATIVE USE The soft wood was used for fuel, but little else. The sap was enjoyed as a chewing gum.

■ LOCAL SITES Common at Cypress Bowl, on Hollyburn Mountain and Grouse Mountain and at Lynn Headwaters Regional Park, often with mountain hemlock and yellow cedar. The tallest Pacific silver fir in B.C. is 43.9 m.

GRAND FIR
Abies grandis • Pine family: *Pinaceae*

■ DESCRIPTION Grand fir is a remarkably fast-growing conifer that reaches heights of over 90 m. Its bark is thin and blistery on young trees, roughened into oblong plates divided by shallow fissures on older ones. The cones are erect, cylindrical, to 10 cm long and green to brown. The needles are dark green and flat, 2-4 cm long, grooved on top, with two white bands of stomata below. This is the tallest of the true firs and was aptly named "grand" by the botanist and explorer David Douglas.

■ HABITAT A low-elevation species on the southern coast, most commonly found in moist areas where it grows with Douglas fir and western red cedar.

■ NATIVE USE The wood was used as a fuel and to make canoes, fishhooks and hand tools. The boughs were brought inside as an air purifier.

■ LOCAL SITES Scattered sitings throughout the forested Lower Mainland. Large old specimens can be seen on the old Marine Drive behind U.B.C. Botanical Gardens. B.C.'s tallest grand fir, 71.3 m, is in Chilliwack River Ecological Reserve.

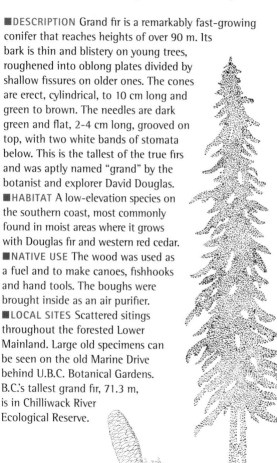

YELLOW CYPRESS or YELLOW CEDAR
Chamaecyparis nootkatensis • Cypress family: *Cupressaceae*

■DESCRIPTION Yellow cypress is a large, slow-growing conifer of conical habit that often exceeds 45 m in height. Its thin, greyish brown bark can be shed in vertical strips. The reddish brown cones are round, to 1-2 cm across, with 4-6 scales tipped with pointed bosses (red cedar has egg-shaped cones). The bluish green leaves are prickly to touch and more pendulous than the red cedar's. Yellow cypress was first documented in Nootka Sound on the west coast of Vancouver Island in 1791 by Archibald Menzies, hence the species name. Its genus name is now being changed to *Cupressus*.

■HABITAT On the southern coast it grows in moist forests at mid to high elevations.

■NATIVE USE The wood was used for carving fine objects such as bentwood boxes, chests and intricately carved canoe paddles. The bark is softer than red cedar's and women used it to make clothing, blankets, baskets, rope and hats.

■LOCAL SITES Can be seen at Cypress Bowl, Hollyburn Mountain and midway up the Grouse Grind. The tallest yellow cypress in B.C. is 62 m.

< *The distinct pendulous profile of the yellow cypress, on the right in this photo (top).*

SITKA SPRUCE
Picea sitchensis • Pine family: *Pinaceae*

■ **DESCRIPTION** Sitka spruce is often seen on rocky outcrops as a twisted dwarf tree, though in favourable conditions it can exceed 90 m in height. Its reddish brown bark is thin and patchy, a good identifier when the branches are too high to observe. The cones are gold brown, to 8 cm long. The needles are dark green, to 3 cm long and sharp to touch. Sitka spruce has the highest strength-to-weight ratio of any B.C. tree. It was used to build the frame of Howard Hughes' infamous plane *Spruce Goose*.

■ **HABITAT** A temperate rainforest tree that does not grow farther than 200 km from the ocean.

■ **NATIVE USE** The new shoots and inner bark were a good source of vitamin C. The best baskets and hats were woven from spruce roots, and the pitch (sap) was often chewed as a gum.

■ **LOCAL SITES** Lynn Headwaters Regional Park, Deer Lake, Noons Creek in Port Moody. Large specimens can be seen at the south end of Swordfern Trail in Pacific Spirit Park and at Third Beach in Stanley Park. The tallest Sitka spruce in the world, 95.8 m, is up Carmanah Creek on Vancouver Island.

SHORE PINE
Pinus contorta var. *contorta* • Pine family: *Pinaceae*

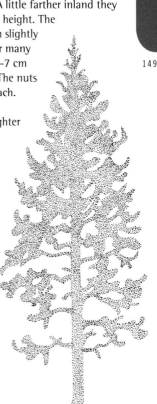

■DESCRIPTION Depending on where they are growing, shore pines vary dramatically in size and shape. By the shoreline they are usually stunted and twisted from harsh winds and nutrient-deficient soil. A little farther inland they can be straight-trunked to 20 m in height. The small cones, 3-5 cm long, are often slightly lopsided and remain on the tree for many years. The dark green needles are 4-7 cm long and grow in bundles of two. The nuts are edible but small and hard to reach. Another variety, lodgepole pine (*P. contorta* var. *latifolia*), grows straighter and taller, to 40 m.

■HABITAT The coastal variety grows in the driest and wettest sites, from low to high elevations.

■NATIVE USE The straight wood was used for teepee poles, torches and arrow and harpoon shafts.

■LOCAL SITES Shoreline trees can be seen from Caulfeild Cove to Horseshoe Bay in West Vancouver. Others are at Camosun Bog, Burns Bog, Mundy Lake and Deer Lake.

WESTERN WHITE PINE
Pinus monticola • Pine family: *Pinaceae*

■ **DESCRIPTION** Western white pine is a medium-size symmetrical conifer to 40 m in height. Its bark is silver grey when young, dark brown and scaly when old. The cones are 15-25 cm long and slightly curved. The bluish green needles are 5-10 cm long and grow in bundles of 5. The species name *monticola* means "growing on mountains."
■ **HABITAT** On the southern coast it grows on moist to wet soils at low to high elevations.
■ **NATIVE USE** The bark was peeled in strips and sewn together with roots to make pine-bark canoes. The pitch was used for waterproofing.
■ **LOCAL SITES** Beautiful mid-size white pines can be seen at Cypress Bowl and First Lake on Hollyburn Mountain.

COMMON JUNIPER
Juniperus communis • Cypress family: *Cupressaceae*

■DESCRIPTION Common juniper is a prostrate conifer that rarely exceeds 1 m in height and 4 m in diameter. Its bluish green needles are very sharp and prick when handled. The light green "berries" (cones) turn a dark bluish black in the second year. These mature "berries" are used to flavour gin. Common juniper is the only circumpolar conifer in the Northern Hemisphere.

■HABITAT Found on dry rocky outcrops from low to alpine elevations, and occasionally in bogs.

■NATIVE USE The wood was only used medicinally, as it rarely attains a size large enough for woodworking or carving.

■LOCAL SITES Open rocky areas from North Vancouver to Lighthouse Park and on to Whistler.

DOUGLAS FIR
Pseudotsuga menziesii • Pine family: *Pinaceae*

■ DESCRIPTION Douglas fir is a tall, fast-growing conifer to 90 m in height. Its bark is thick, corky and deeply furrowed. The ovate cones are 7-10 cm long and have 3 forked bracts protruding from the scales; the cones hang down from the branches, unlike true firs' cones, which stand up. The needles are 2-3 cm long, pointed at the apex, with a slight groove on the top and two white bands of stomata on the underside. The common name commemorates the botanist and explorer David Douglas.

■ HABITAT Can tolerate dry to moist conditions from low to high elevations. Reaches its tallest size near the coast.

■ NATIVE USE The wood was used for teepee poles, smoking racks, spear shafts, fishhooks and firewood.

■ LOCAL SITES Common. Large specimens can be seen on Boom Trail in Pacific Spirit Park and at Hollyburn Mountain, Lighthouse Park, Stanley Park, Capilano Regional Park, Chilliwack River and Lynn Headwaters Regional Park. The tallest Douglas fir in Canada is 94.3 m.

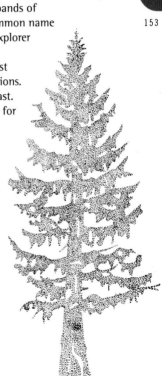

WESTERN RED CEDAR
Thuja plicata • Cypress family: *Cupressaceae*

■DESCRIPTION Western red cedar is a large, fast-growing conifer with heights exceeding 60 m. Its bark sheds vertically and ranges from cinnamon red on young trees to grey on mature ones. The bases of older trees are usually heavily flared, with deep furrows. The egg-shaped cones are 1 cm long, green when young, turning brown and upright when mature (yellow cypress has round cones). The bright green leaves are scale-like, with an overlapping-shingle appearance.

Western red cedar is B.C.'s provincial tree. On old stumps, springboard marks can be seen 2-3 m above the ground; these allowed early fallers to get away from the flared bases. The shingle industry is now the biggest user of red cedar.

■HABITAT Thrives on moist ground at low elevations. Will tolerate drier or higher sites but will not attain gigantic proportions.

■NATIVE USE First Nations people know this tree as "the tree of life." It supplied them from birth to death, from cradle to coffin. The wood was used to make dugout canoes, fishing floats, paddles, bowls, masks, totem poles, ornamental boxes and spear and arrow shafts. The bark was shredded for clothing, diapers, mats, blankets, baskets and medicine.

■LOCAL SITES Common in forested areas at mid to low elevations. Huge cedars can be seen at Third Beach in Stanley Park (4.3 m in diameter) and Lynn Headwaters Regional Park (4 m). The tallest western red cedar in B.C. is 59.2 m.

WESTERN HEMLOCK
Tsuga heterophylla • Pine family: *Pinaceae*

CONIFERS

■ DESCRIPTION Western hemlock is a fast-growing pyramidal conifer to 60 m in height. Its reddish brown bark becomes thick and deeply furrowed on mature trees. The plentiful cones are small (2-2.5 cm long), conical and reddish when young. The flat, light green leaves vary in size from 0.5 to 2 cm long. The main leaders and new shoots are nodding, giving the tree a soft, pendulous appearance that is good for identification. Western hemlock is the state tree of Washington.

■ HABITAT Flourishes on the Pacific coast from Alaska to Oregon and from low levels to 1,000 m, where it is replaced by mountain hemlock (*T. mertensiana*).

■ NATIVE USE The wood has long been used for spear shafts, spoons, dishes, roasting spits and ridgepoles. The bark was boiled to make a red dye for wool and basket material.

■ LOCAL SITES Common in forested areas. Large specimens can be seen at Lynn Headwaters Regional Park. B.C.'s tallest hemlock, 75.6 m, is near Tahsish River on the west coast of Vancouver Island.

MOUNTAIN HEMLOCK
Tsuga mertensiana • Pine family: *Pinaceae*

■**DESCRIPTION** Mountain hemlock is a smaller, stiffer tree than western hemlock (*T. heterophylla*; see page 157). It is often stunted by its harsh environment, but with good conditions it can slowly grow to 40 m in height. The reddish brown bark is rough and deeply ridged, even on young trees. The cones are up to 8 cm long, compared with 2.5 cm for the western hemlock. The bluish green leaves are equal lengths to 2-5 cm long. The species name commemorates the German botanist Franz Karl Mertens.

■**HABITAT** Typically found at higher elevations, associating with Pacific silver fir (*Abies amabilis*; see page 141) and sub-alpine fir (*A. lasiocarpa*). Well-adapted to the short growing season and heavy snow packs.

■**LOCAL SITES** Top of the Grouse Grind and from Cypress Bowl to Whistler Mountain. Does not naturally grow at lower elevations.

WESTERN YEW or PACIFIC YEW
Taxus brevifolia • Yew family: *Taxaceae*

■**DESCRIPTION** Western yew is a small conifer from 3 to 15 m in height. It is usually seen as a straggly shrub or small tree in the understory of larger trees. Its thin brownish bark is scale-like, exposing reddish purple patches that distinguish it from the European species. Female trees produce a beautiful but poisonous red berry that ripens in August/September. The flat needles are 2-3 cm long, dark green above with white bands below. The cancer-fighting drug Taxol is extracted from yew bark.

CAUTION: the berries are considered poisonous.

■**HABITAT** Found intermittently on a variety of forested sites at low to mid elevations.

■**NATIVE USE** Western yew was considered the best wood for making bows.

■**LOCAL SITES** Smaller trees are found scattered through lower forested areas. Large specimens can be seen at Capilano Salmon Hatchery, Locarno Beach and Spanish Banks East concession stands. The tallest yew in B.C., 22.25 m, is at Fulford Harbour on Saltspring Island.

GLOSSARY

Achene	A small, dry, one-seeded fruit (e.g., sunflower seeds).
Anther	The pollen-bearing (top) portion of the stamen.
Axil	The angle made between a stalk and a stem on which it is growing.
Biennial	Completing its life cycle in two growing seasons.
Boss	Knob-like studs, as in the points on cones of yellow cypress.
Bract	A modified leaf below the flower.
Catkin	A spike-like or drooping flower cluster, male or female (e.g., cottonwood).
Corm	An underground swollen stem capable of producing roots, leaves and flowers.
Deciduous	A plant that sheds its leaves annually, usually in the autumn.
Dioecious	Male and female flowers on separate plants.
Epiphyte	A plant that grows on another plant for physical support, without robbing the host of nutrients.
Herbaceous perennial	A nonwoody plant that dies back to the ground each year and regrows the following season.
Lenticel	Raised organs that replace stomata on a stem.
Node	The place on a stem where the leaves and auxiliary buds are attached.
Obovate	Oval in shape, with the narrower end pointing downward, like an upside down egg.
Panicle	A branched inflorescence.
Petiole	The stalk of a leaf.
Pinnate	A compound leaf with the leaflets arranged on both sides of a central axis.
Pinnule	Leaflet of a pinnately compound leaf.

GLOSSARY

Rhizome	An underground modified stem. Runners and stolons are on top of the ground.
Scape	A leafless stem rising from the ground. It may support one or many flowers.
Sepal	The outer parts of a flower, usually green.
Sori	Spore cases.
Stipe	Stalk (petiole), usually referring to ferns.
Stolon	A stem or branch that runs along the surface of the ground and takes root at the nodes or apex, forming new plants.
Stomata	The pores in the epidermis of leaves, usually seen as white.
Style	The stem of the pistil (female organ).

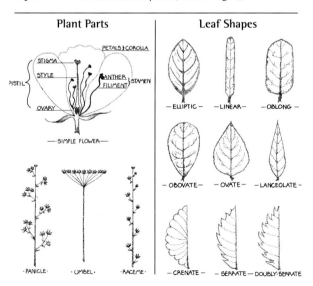

BIBLIOGRAPHY

Adolph, Val. *Tales of the Trees*. Key Books, Delta, B.C., 2000

Burns, Bill. *Discover Burns Bog*. Hurricane Press, Vancouver, B.C., 1997

Clark, Lewis J. *Wild Flowers of British Columbia*. Gray's Publishing Ltd., Sidney, B.C., 1973

Craighead, John J., Frank C. Craighead, Jr., and Ray J. Davis. *A Field Guide to Rocky Mountain Wildflowers*. Houghton Mifflin Company, Boston, Massachusetts, 1963

Haskin, L. L. *Wild Flowers of the Pacific Coast*. Binford and Mort, Portland, Oregon, 1934 (republished 1977, Dover Publications, New York)

Klinka, K., V. J. Krajina, A. Ceska and A. M. Scagel. *Indicator Plants of Coastal British Columbia*. University of British Columbia Press, Vancouver, B.C., 1989

Lyons, C. P. *Trees, Shrubs and Flowers to Know in British Columbia*. J. M. Dent and Sons, Toronto, Ontario and Vancouver, B.C., 1976 (1st ed. 1952)

Pojar, Jim and MacKinnon, Andy. *Plants of the Pacific Northwest Coast*. Lone Pine Publishing, Vancouver, B.C. 1994

Sargent, Charles Sprague. *Manual of the Trees of North America*. Dover Publications, New York, 1965, two volumes (originally published in 1905 by Houghton Mifflin Company, Boston, Massachusetts)

Smith, Kathleen M. and Nancy J. Anderson. *Nature West Coast As Seen in Lighthouse Park*. Sono Nis Press, Victoria, B.C., 1988

Stoltmann, Randy. *Hiking Guide to the Big Trees of Southwestern British Columbia*. Western Canada Wilderness Committee, Vancouver, B.C., 1987

Szczawinski, A. F., and George A. Hardy. *Guide to Common Edible Plants of British Columbia*. Handbook No. 20, Royal British Columbia Museum, Victoria, B.C., 1974

Turner, Nancy J. *Plants in British Columbia Indian Technology*. Handbook No. 38, Royal British Columbia Museum, Victoria, B.C., 1979

INDEX

Abies,
   amabilis, 141, 158
   grandis, 143
Acer,
   circinatum, 119
   macrophyllum, 121
Achillea millefolium, 19
Achlys triphylla, 63
Adiantum pedatum, 81
Agassiz, 133
Alder,
   red, 123
   Sitka, 124
Allium cernuum, 32
Alnus,
   rubra, 123
   sinuata, 124
Alouette Lake, 12, 38, 39, 59, 76, 94, 104, 112, 121
alumroot, small-flowered, 23
Amelanchier alnifolia, 70
Anaphalis margaritacea, 21
Aquilegia formosa, 51
Arbutus, 127
Arbutus menziesii, 127
Arctostaphylos uva-ursi, 75
Aruncus dioicus, 103
Athyrium filix-femina, 79, 84
avens, large-leafed, 41

Bear Mountain, 28, 75, 101
bearberry, 75
Beaver Lake, 57, 73
Betula papyrifera, 129
bittersweet, European, 66
blackberry,
   cutleaf, 69
   Himalayan, 69
   trailing, 69
Blackcomb, 116
Blechnum spicant, 82
bleeding heart, Pacific, 52
blueberry, oval-leafed, 73

bog laurel, western, 108
Boom Trail (Pacific Spirit Park), 42, 64, 91, 96, 153
Boundary Bay, 13, 19
Bowen Island, 51
Boykinia elata, 22
Boykinia, coast, 22
Brandywine Falls, 29, 112
Brodiaea,
   coronaria, 27, 30
   hyacinthina, 27
broom, Scotch, 110
bunchberry, 55
Burnaby, 26, 105
Burnaby Mountain Park, 62
Burns Bog, 9, 53, 55, 57, 90, 97, 99, 106, 108, 109, 113, 115, 149
buttercup, creeping, 50

Camosun Bog, 53, 73, 90, 108, 109, 149
Campbell Valley Regional Park, 71
canoe birch, 129
Capilano Regional Park, 153
Capilano Salmon Hatchery, 159
cascara, 136
cat-tail, 91
Caulfeild Cove, 11, 16, 27, 30, 32, 33, 34, 37, 70, 93, 106, 149
Cedar,
   western red, 155
   yellow, 145
Central Park, Burnaby, 26, 35, 41, 105
Chamaecyparis nootkatensis, 145
cherry, bitter, 129, 135
Chilliwack, 20, 58, 133
Chilliwack River, 143, 153
Chrysanthemum leucanthemum, 15
Cirsium vulgare, 13
Claytonia sibirica, 60
Clintonia uniflora, 29
clubmoss, running, 92
coltsfoot, 17
columbine, red, 51

163

INDEX

Cornus,
  canadensis, 55
  nuttallii, 55, 131
  stolonifera, 113
cottonwood, 133
cow-parsnip, 45
crab apple, Pacific, 125
Crataegus douglasii, 139
Cryptogramma crispa, 83
currant, red flowering, 111
Cypress Bowl, 15, 28, 29, 31, 39, 92, 95, 141, 145, 150, 158
Cypress Park, 12, 111,117
cypress, yellow, 145
Cytisus scoparius, 110

daisy, oxeye, 15
Daucus carota, 46
death camas, 33
Debouville Slough, 104, 113
deer cabbage, 95
Deer Lake Park, 24, 35, 55, 62, 106, 109, 115, 129, 147, 149
devil's club, 112
Dicentra formosa, 52
Digitalis purpurea, 44
Dogwood,
  Pacific, 55, 131
  red-osier, 113
  dwarf, 55
Douglas fir, 153
Drosera rotundifolia, 90
Dryopteris expansa, 79, 84
Dunbar (beach), 43, 101

Epilobium angustifolium, 54
Equisetum,
  arvense, 89, 95
  hyemale, 96
Erythronium oregonum, 38

false azalea, 107
falsebox, 114

Fauria crista-galli, 95
fern,
  bracken, 86
  deer, 82
  lady, 79, 84
  licorice, 82, 87, 121
  maidenhair, 81
  mountain, 83
  parsley, 83
  shield, 84
  spiny wood, 79, 84
fir,
  grand, 143
  Pacific silver, 141, 158
fireweed, 54
First Lake, 29, 31, 95, 150
foam flower, 26
fool's huckleberry, 107
fool's onion, 27, 30
foxglove, 44
Fraser River, 47, 64, 66, 83, 91, 96, 98, 113, 117, 139
Fraser Valley, 11, 13, 42, 114, 131
fringecup, 24, 41
Fritillaria camschatcensis, 34
Fulford Harbour, 159

gale, sweet, 115
Garibaldi, 17, 112
Gaultheria shallon, 76
Geranium robertianum, 62
Geum macrophyllum, 41
goat's beard, 103
Golden Ears Provincial Park, 26, 37, 51, 63
goldenrod, Canada, 20
Goodyera oblongifolia, 61
Grindelia integrifolia, 16
Grouse Grind, 12, 31, 92, 116, 138, 145, 158
Grouse Mountain, 39, 55, 124, 141
gumweed, 16

hardhack, 99, 115
Harrison Hot Springs, 86
Harrison Lake, 28, 37, 39, 41, 46, 61, 75, 76, 101, 104, 105, 113, 117, 121, 129, 133, 138
Harrison River, 81, 112, 125
hawthorn, black, 34, 139
heal-all, 49
hedge-nettle, Cooley's, 48
hemlock,
 mountain, 158
 western, 157
Heracleum lanatum, 45
herb robert, 62
Heuchera micrantha, 23
Hollyburn Mountain, 12, 28, 29, 31, 39, 55, 73, 95, 111, 116, 138, 141, 145, 150, 153
Holodiscus discolor, 101
honeysuckle, climbing, 11
Horseshoe Bay, 11, 16, 23, 34, 46, 70, 149
horsetail, common, 89
huckleberry, red, 77

Indian celery, 45
Indian hellebore, 31, 112
Indian pipe, 94
Indian plum, 102
Iris pseudacorus, 64

Jericho Beach Park, 18, 42, 64, 66, 91,101
juniper, common, 151
Juniperus communis, 151

Kalmia microphylla ssp. *occidentalis*, 108
kinnikinnick, 75
Kitsilano Beach, 45, 66
knotweed, giant, 58

Labrador tea, 108, 109

Lactuca muralis, 14
Ledum Groenlandicum, 108, 109
lettuce, wall, 14
Lighthouse Park, 12, 22, 23, 27, 28, 30, 37, 38, 39, 42, 59, 61, 70, 75, 76, 78, 81, 87, 97, 104, 110, 111, 117, 124, 125, 127, 131, 136, 151, 153
Lilium columbianum, 37
lily,
 bead, 29
 black, 34
 corn, 31
 harvest, 37, 30
 of the valley, false, 35
 tiger, 37
 western white fawn, 38
Linnaea borealis, 12
Lion's Bay, 23, 127
Locarno Beach, 17, 129, 159
Lonicera,
 ciliosa, 11
 involucrata, 106
loosestrife, purple, 65
Lycopodium clavatum, 92
Lynn Headwaters Regional Park, 60, 77, 86, 141, 147, 153, 155, 157
Lysichiton americanum, 57
Lythrum salicaria, 65

madrone, Pacific, 127
Mahonia nervosa, 78
Maianthemum dilatatum, 35
Malus fusca, 125
maple,
 bigleaf, 119, 121
 vine, 119
Menziesia ferruginea, 107
Mimulus guttatus, 43
miner's lettuce, Siberian, 60
Minnekhada Regional Park, 19, 74, 106, 117, 125, 139
monkey-flower, yellow, 43
Monotropa uniflora, 94

INDEX

Mount Seymour Provincial Park, 60, 75
mountain ash, Sitka, 138
mountain heather, pink, 116
mountain lover, 114
Mundy Lake Park, 35, 39, 57, 73, 74, 90, 94, 108, 109,125, 129, 149
Musqueam Reserve, 19, 113
Myrica gale, 115

nettle, stinging, 93
ninebark, 117
Noons Creek, 34, 35, 55, 97,106,107,113, 125, 147
North Vancouver, 22, 92, 114, 151
northern rice root, 34

oceanspray, 101
Oemleria cerasiformis, 102
Oenanthe sarmentosa, 47
onion, nodding, 32
Oplopanax horridus, 112
Oregon grape, 78

Pachistima myrsinites, 114
Pacific Spirit Park, 11, 24, 35, 39, 42, 52, 59, 60, 64, 74, 78, 91, 93, 96, 101, 105, 113, 121, 123, 129, 135, 136,147, 153
Pacific yew, 159
paper birch, 129
pearly everlasting, 21
Petasites palmatus, 17
Phyllodoce empetriformis, 116
Physocarpus capitatus, 117
Picea sitchensis, 147
piggy-back plant, 25
pine,
    shore, 149
    western white, 150
Pinus,
    contorta, 149
    monticola, 150
Polygonum sachalinense, 58

Polypodium glycyrrhiza, 82, 87
Polystichum munitum, 82, 84
Populus balsamifera spp. trichocarp, 133
Port Coquitlam, 112
Port Moody, 34, 35, 55, 97, 106,107, 113, 125, 147
Potentilla anserina ssp. pacifica, 42
Prospect Point, 45, 119
Prunella vulgaris, 49
Prunus emarginata, 129, 135
Pseudotsuga menziesii, 153
Pteridium aquilinum, 86

queen anne's lace, 46
queen's cup, 29

Ranunculus repens, 50
rattlesnake plantain, 61
red elderberry, 104
red-berried elder, 104
Rhamnus purshiana, 136
Ribes sanguineum, 111
Richmond Nature Park, 24
Rosa,
    gymnocarpa, 98
    Nutkana, 97
rose,
    baldhip, 98
    Nootka, 97, 98
    woodland, 98
Rubus,
    discolor, 69
    laciniatus, 69
    parviflorus, 71
    spectabilis, 74
    ursinus, 69

salal, 76
Salix, lucida ssp. Lasiandra, 137
Salix, scouleriana, 137
salmonberry, 74
Sambucus racemosa, 104

INDEX

Saskatoon berry, 70
scouring rush, 96
self-heal, 49
serviceberry, 70
Shannon Falls, 23, 55, 66, 71, 75, 77, 86, 97, 98, 106, 124
silverweed, 42
skunk cabbage, 57
snowberry, 105
Solanum dulcamara, 66
Solidago canadensis, 20
Sorbus sitchensis, 138
Spanish Banks, 159
Spiraea douglasii, 99, 115
spruce, Sitka, 147
Squamish, 58, 66, 107, 131
Stachys cooleyae, 48
Stanley Park, 7, 11, 22, 23, 35, 41, 45, 70, 71, 104, 106, 111, 119, 121, 123, 125, 135, 147, 153, 155
star flower, western, 53
Stawamus Chief, 107
steeplebush, 99
Streptopus amplexifolius, 28
sundew, round-leafed, 90
sword fern, western, 82, 84
Swordfern Trail (Pacific Spirit Park), 11, 147
Symphoricarpos albus, 105

Tanacetum vulgare, 18
tansy, common, 18
Tahsish River, 157
Taxus brevifolia, 159
Tellima grandiflora, 24, 41
thimbleberry, 71
Third Beach, 119, 123, 147, 155
thistle, bull, 13
Thuja plicata, 155
Tiarella trifoliata, 25
Tolmiea menziesii, 24, 25
Trafalgar (beach), 43, 45, 66, 101
Trientalis latifolia, 53

Trillium, ovatum, 39
Trillium, western, 39
Tsuga,
  heterophylla, 157
  mertensiana, 158
twinberry, black, 106
twinflower, 12
twisted stalk, 28
Typha latifolia, 91

UBC, 17, 143
Urtica dioica, 93

Vaccinium,
  ovalifolium, 73
  parvifolium, 77
Vancouver Island, 27, 30, 110, 127, 147, 157
Vancouver Maritime Museum, 45
vanilla leaf, 63
Veratrum viride, 31
Viola sempervirens, 59
violet, evergreen, 59

water-parsley, Pacific, 47
West Vancouver, 22, 92, 98, 114, 127, 131, 149
Whistler, 17, 20, 26, 51, 61, 83, 107, 112, 116, 138, 151, 158
Whytecliffe Park, 27, 30, 32, 33, 62, 70, 127
wild carrot, 46
wild hyacinth, 30
willow,
  Pacific, 137
  Scouler's, 137
Wreck Beach, 17, 43, 101

yarrow, 19
yellow flag, 64
yew, western, 159

Zygadenus venenosus, 33

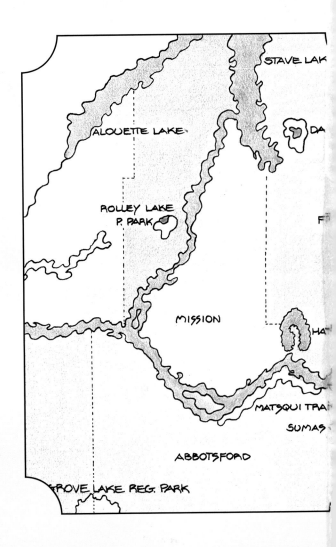